T5-AGA-660

Engineering Robust Designs with Six Sigma

Engineering Robust Designs with Six Sigma

John X. Wang

Upper Saddle River, NJ • Boston • Indianapolis • San Francisco
New York • Toronto • Montreal • London • Munich • Paris • Madrid
Capetown • Sydney • Tokyo • Singapore • Mexico City

The publisher offers excellent discounts on this book when ordered in quantity for bulk purchases or special sales, which may include electronic versions and/or custom covers and content particular to your business, training goals, marketing focus, and branding interests. For more information, please contact:

U. S. Corporate and Government Sales
(800) 382-3419
corpsales@pearsontechgroup.com

For sales outside the United States, please contact:

International Sales
international@pearsoned.com

Visit us on the Web: www.phptr.com

Library of Congress Cataloging-in-Publication Data

Wang, John X., 1962–
Engineering robust designs with Six Sigma / John X. Wang.
p. cm.
Includes bibliographical references and index.
ISBN 0-13-144855-2 (hardcover: alk. paper)
1. Quality control—Statistical methods. 2. Experimental design. I. Title.

MIL TS156.W335 2005
658.5'752--dc22

2004028711

ISBN 0-13-144855-2
Text printed in the United States on recycled paper at Courier in Westford, Massachusetts.
First printing, March 2005

To my wife Xin and my sons Sonny and Cheney for your love and support.

To my parents who encouraged me to write this new book.

Contents

Preface

This book is written for engineers who continue to have a dramatic impact on our world. The *essence* of the engineer's job is the engineering of robust products for the benefit of all people. A robust product is not just strong; it is also efficient, flexible, mistake-proof, and affordable. The term *engineering robust products* encompasses design and process flexibility, which rapidly and affordably accommodate changing customer requirements. As an engineer and an engineering manager, I have worked for the power utility, aerospace, transportation, automotive, and appliance industries. I have had the privilege to witness the essential needs and challenges of developing dependable power supplies, reliable manned-flight vehicles, low-emission transportation systems, high-efficiency cars, and cost-competitive appliances.

Engineers need to consider many factors when developing new products. For example, in developing a hybrid car, engineers determine precisely what function it needs to perform; design and test the components; fit the components together in an integrated plan; and evaluate the design's overall effectiveness, cost, reliability, serviceability, and safety. This process applies to many different products such as magnetic resonance imaging equipment, the Internet, gas turbines, helicopters, and even toy gyrocopters. Within the next decade, emerging technologies will make possible novel applications that integrate computation into design, manufacturing, and application environments: smart materials, self-reconfiguring robots, self-assembling nanostructures. Companies often face the challenge of achieving coherent and robust behavior from the interactions of a multitude of components and their interaction with design, manufacturing, and application. The new environments highlight the limits of current engineering practices and techniques, which are still highly dependent on precision parts and controlled processes to make products dependable and reliable.

The goal of engineering robust products is to deliver what customers expect at an affordable cost regardless of usage; degradation over product life; and variation in manufacturing, suppliers, distribution, delivery, installation, and service. The use of Dr. Genichi Taguchi's robust design principles (1987) allows experiments to be performed and prototypes to be tested on multiple factors at once so that products become insensitive to usage conditions and other uncontrollable factors. Because randomness and scatter are a part of reality everywhere, Six Sigma (SS) techniques are becoming necessary to design quality into products. *Six Sigma quality* describes a highly robust product development and manufacturing process. Achieving SS quality ensures both dependable products and production efficiency.

This book's title, *Engineering Robust Products with Six Sigma,* emphasizes the use of a disciplined process in conjunction with a robust product design. The appropriate application of robust engineering principles with the Six Sigma process enables development programs to quickly deliver high-quality, low-cost products that fully meet customers' needs. Product value attracts customers, quality brings respect, and innovation distinguishes one product from its competition.

The use of Six Sigma to engineer robust products bridges the gap between research and development, invention and innovation, and product and/or process development. SS enables engineers to develop products and processes that will perform consistently as intended under a wide range of conditions throughout their life cycle. Using the Six Sigma techniques also allows engineers to develop or change product formulas and process settings to achieve desired performance at the lowest cost and in the shortest time.

A few of the book's reviewers work or have worked for the National Aeronautics and Space Administration. This reminds me of these facts: "All the astronauts who landed on the Moon were *engineers*," and we are following in their footprints when engineering robust products for humankind.

ACKNOWLEDGMENTS

The author would like to thank all the reviewers who have helped to refine this book with their constructive comments. My special thanks go to Dr. Donald W. Sova and Mr. Jeffrey B. Lynch for their extended support while they reviewed the entire manuscript.

About the Author

John X. Wang is a Manager of Reliability Engineering at Maytag Corporation, where he taught the Design for Six Sigma training courses as a Manager of Design for Six Sigma. Dr. Wang has been Six Sigma Master Black Belt certified by the Visteon Corporation and Six-Sigma Black Belt certified by the General Electric Company. The author or coauthor of numerous professional papers on Six Sigma, fault diagnosis, reliability engineering, and other topics, Dr. Wang has been a Certified Reliability Engineer of the American Society for Quality and a Six Sigma Master Black Belt certified by the International Quality Federation. He received B.A. (1985) and M.S. (1987) degrees from Tsinghua University, Beijing, China; and a Ph.D. degree (1995) from the University of Maryland at College Park. Dr. Wang lives in Marion, Iowa, with his wife and two sons.

1 Achieving Robust Designs with Six Sigma
Dependable, Reliable, and Affordable

Developing "best-in-class" robust designs is crucial for creating competitive advantages. Customers want their products to be dependable—"plug-and-play." They also expect them to be reliable—"last a long time." Furthermore, customers are cost-sensible; they anticipate that products will be affordable. Becoming *robust* means seeking win–win solutions for productivity and quality improvement. So far, robust design has been a "road less traveled." Very few engineering managers and professionals are aware of robust design methods; even fewer of them have hands-on experience in developing robust designs. As a breakthrough philosophy, process, and methodology, Six Sigma offers a refreshing approach to systematically implement robust designs. This chapter outlines a process for engineering robust designs with Six Sigma and provides a road map.

1.1 Six Sigma and Robust Design

Six Sigma is a rigorous and disciplined methodology that uses data and statistical analysis to measure and improve a company's operational performance. It identifies and eliminates "defects" in product development, manufacturing, and service-related processes. The goal of Six Sigma is to increase profits by eliminating variability, defects, and waste that undermine customer loyalty.

A *best-in-class* robust design starts with three categories of static response metrics: the smaller-the-better, the nominal-the-best, and the larger-the-better. Each of these characteristics should be measurable on a continuous scale.

- A *smaller-the-better* response is a measured characteristic with an ideal value of zero. As the value for this type of response decreases, quality improves.
- A *nominal-the-best* response is a measured characteristic with a specific target (nominal) value that is considered ideal.
- A *larger-the-better* response is a measured characteristic with an ideal value of infinity. As the value for this type of response increases, quality improves.

Besides static responses, dynamic responses are also encountered when developing engineering products. A *dynamic response* is a characteristic that, ideally, increases along a continuous scale in proportion to input from the system. Dynamic responses should be related to the transfer of energy through the system. To develop robust products, dynamic formulations are recommended for the maximum benefit of the application of a Parameter Design methodology (see Chapter 7). Using a dynamic response provides the greatest long-term benefits, but it requires the most engineering know-how.

The Six Sigma approach for engineering robust designs depends heavily on formulating the Voice-of-Customers (VOCs) and Critical-to-Quality (CTQ) characteristics through experiments. The following steps provide a thorough, organized framework for planning, managing, conducting, and analyzing robust design experiments.

1. Identify project and organize team
2. Develop VOC models
3. Formulate the CTQs based on VOCs
4. Control the energy transformation for each CTQ
5. Determine control and noise factors for each CTQ
6. Establish the control factor matrix

These steps, although specified sequentially, should not be used as a "cookbook approach" to experimentation; instead, they should be used in an iterative way. During each stage of development, consider the decisions that were made in earlier steps. Your team may need to revisit previous steps in light of insights gained farther along in the process.

1.2 IDENTIFY PROJECT AND ORGANIZE TEAM

Six Sigma provides a product development team with the tools to improve design capability. The first step in the robust design process is to identify the project (Figure 1-1) and organize a team. This section shows how to develop the project-selection criteria and successful project characteristics.

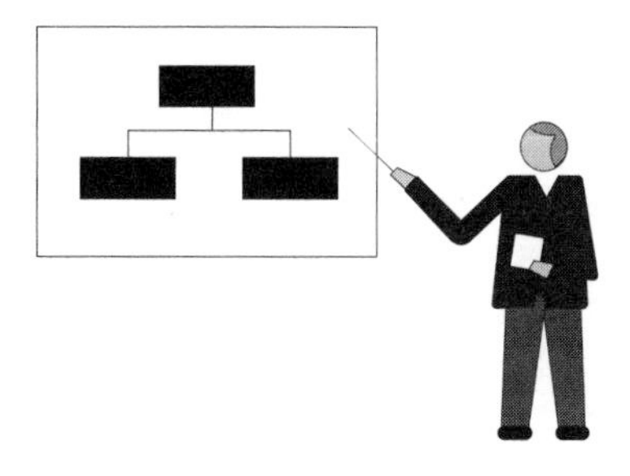

Figure 1-1 Identify a robust design project.

As with any project, effective planning and selection of the right team can make the difference between success and failure. Project selection should be based on the potential for increasing customer satisfaction, increasing reliability, incorporating new technology, reducing cost, reducing warranty service, and achieving best-in-class.

The characteristics of a successful project are (1) a clear objective, including the desired outcome; and (2) a cross-functional team that includes suppliers, thorough planning, and management support. Management sponsorship and support is critical for team success. Management's role is to provide necessary resources, empower the team, and remove obstacles to progress.

1.3 DEVELOP VOC MODELS

The *Voice-of-Customer* process is used to capture the requirements and/or feedback from customers (internal or external) to provide them with the best-in-class service or product quality. This process is all about being proactive and constantly innovative in order to capture the changing requirements of customers over time. The second step in the robust design process is to begin to develop VOC models. The following are several key requirements and inputs of a Voice-of-Customer model:

- Perceived result
- Customer intent
- Customer and engineering metrics
- Intended function
- Response criteria

Customer requirements are the starting point for determining what to measure in an experiment. But, customer performance metrics are sometimes vague, usually subjective, and typically expressed in nontechnical

- ➢ Identify customer needs
- ➢ Perform surveys
- ➢ Communicate with focus groups
- ➢ Turn customer data into VOC models

Figure 1-2 Establish Voice-of-Customer models.

terms. So, to produce quality products, the engineer must translate customer performance metrics into measurable, objective engineering metrics (Figure 1-2).

For example, the VOCs convey to the engineer what customers want and how they perceive what they actually get. The Voice-of-Customer model is the engineers' interpretation, in engineering terms and functions, of customers' *perceived result.* But the perceived result is a subjective perception of what customers get from the product. Together, these represent the customer's world. When the perceived result doesn't match her or his voice, the customer is disappointed.

Traditionally, the mismatch between the voice of the customer and the perceived result has been addressed by attempting to "solve the problem" when it becomes evident. It would be preferable, however, to anticipate customers' expectations and design products to meet them proactively.

The term *Voice-of-Customer* is used to describe the stated and unstated needs or requirements of the customer, and there are a variety of ways to capture VOCs:

- Direct discussion or interviews
- Surveys

- Focus groups
- Customer specifications
- Observation
- Warranty data
- Field reports
- Complaint logs
- And so on . . .

To design robust products, the engineer must determine which system or subsystem to study and establish technical metrics that quantify the system's ability to satisfy the VOC. The Voice-of-Customer model ultimately determines what is critical to quality—the focus of the next section.

1.4 FORMULATE CRITICAL-TO-QUALITY CHARACTERISTICS

CTQs (*Critical-to-Quality*) are the key measurable characteristics of a product or process whose performance standards or specification limits must be met to satisfy the VOCs (Figure 1-3). They align improvement or design efforts with customer requirements.

➢ *Metric* means measurement
➢ Measure product quality levels
➢ Help understand design tradeoffs
➢ Develop based on VOC modeling

Figure 1-3 CTQ: An engineering metric.

CTQs represent the product or service characteristics defined by the customer (internal or external). They can include the upper- and lower-specification limits or any other factors related to the product or service. A CTQ item usually must be determined from a qualitative customer statement and "translated" into an actionable, quantitative business specification.

In robust design, a Critical-to-Quality characteristic should be related to the perceived result. In a robust design experiment, the measured output of the system is the CTQ. The CTQ performance is frequently inconsistent with the ideal perceived result because of noise. When determining what to measure in an experiment—the CTQ—first consider the customer's perspective of system functionality.

For example, customers use their brakes with the intent of slowing down or stopping the car. The ideal perceived result is for the car to smoothly slow down or come to a stop every time the brakes are applied. With this in mind, the team must now determine which portion of the brake system to focus on and establish a CTQ, in engineering metrics, that quantifies the functionality of that system.

For nonrobust brake systems, the perceived result is often conveyed in terms that describe unintended results, such as noisy or rough. Optimizing a CTQ related to brake functionality—the distance to stop—minimizes unintended results. This philosophy represents a change in the way the engineer approaches the design process.

In sum, a CTQ is an engineering metric that quantifies the system's functionality (i.e., its ability to meet the VOCs). Critical-to-Quality characteristics drive efforts to control energy transformation within a product or system.

1.5 CONTROL ENERGY TRANSFORMATION FOR EACH CTQ CHARACTERISTIC

To fulfill the intent of the system, the customer does something that initiates a transfer of energy, which produces a CTQ that might be categorized as either the intended result or the unintended result (error state). Because energy transfer creates CTQs, the system must be studied in terms of this transfer. Such a study will help the team identify a response that quantifies the system's production of intended results. It is important to consider the following:

- Energy can neither be created nor destroyed.
- Energy can be transformed into various states.
- Only one energy state is intended, or ideal.

Maximizing the amount of energy used to produce an intended result will minimize the amount available to produce unintended results, or error states (Figure 1-4). Robust design shifts from examining the error states and searching for remedies, to studying the functional intent of the system and exploring ways of optimizing it.

> ➢ Energy: Mechanical, thermal, electrical, chemical, ...
> ➢ Target or ideal state: 100 percent energy utilization
> ➢ Avoid error states; that is, energy is transferred smoothly

Figure 1-4 Control energy transformation effectively.

Engineering robust products with Six Sigma requires a shift from measuring the symptoms of poor quality to measuring the transformation of energy. This philosophy requires a shift in thinking by the engineer. Robust design enhances quality through a focus on optimizing the system's intended functions—the efficient transfer of all energy.

To depict the intended function in terms of engineering metrics, study the underlying physics of a system, which should yield an engineering metric that quantifies the amount of energy used to produce a result. Use this metric as the Critical-to-Quality characteristic. Maximizing such a CTQ will optimize system functionality.

Clearly, CTQ characteristics depend on the system chosen for study. Many systems are composed of several subsystems and related processes, each with its own intended function. Therefore, what to study must be determined before the team can identify the transfer function and the related CTQs.

The following are the three types of metrics commonly used in industry:

- *Customer metrics*—usually subjective and expressed in nontechnical terms
- *Management metrics*—typically related to productivity or economics
- *Engineering metrics*—quantitative, objective, and physics-based

All of these metrics have their place in the development of quality products and processes. In experimentation though, engineering metrics will provide more useful and reproducible information than either management or customer metrics.

Levels are the different settings a factor can have. For example, if you want to determine how the response (speed of data transmittal) is affected by the factor (connection type), you need to set the factor at different levels (e.g., modem and local area network).

EXAMPLE 1.1

Consider a robust design experiment with the objective to reduce the production of defective donuts. Suppose the management metric yield is the CTQ and the factors (time and temperature) are both tested at two levels.

Yield is the percentage of a product that is free of defects (i.e., the percentage of defect-free products over the total number of products produced). At the low temperature, B_1, increasing time (from A_1 to A_2) increases yield; whereas, at the high temperature, B_2, increasing time (from A_1 to A_2) decreases yield.

A: Time	B: Temperature
A_1 = 5 minutes	B_1 = 200°F
A_2 = 8 minutes	B_2 = 260°F

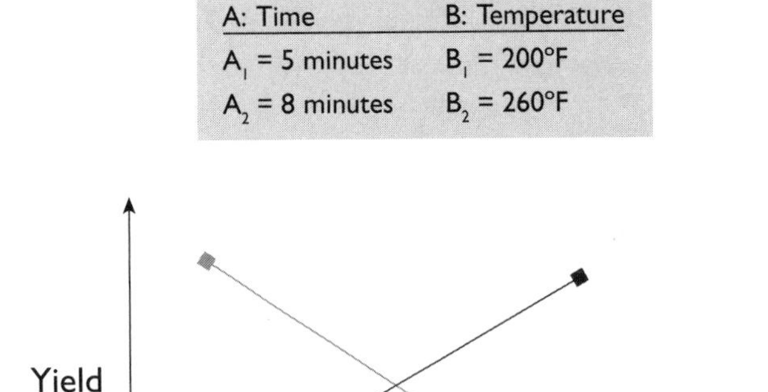

Figure 1-5 Time and temperature interact with respect to yield.

So, the effect of cook time (on yield) depends on the temperature level. This implies that cook time and temperature interact, as is indicated by the nonparallel nature of the lines on the interaction plot of level combinations (see Figure 1-5). An interaction occurs when the response achieved by one factor depends on the level of the other factor. On the interaction plot, when lines are not parallel, there's an interaction.

Suppose, instead, that the CTQ is defined as "color." With color as the CTQ, increasing cook time increases color at either temperature, and increasing temperature increases color at either level of cook time. An interaction occurs when the response achieved by one factor depends on the level of the other factor. On the interaction plot, when lines are not parallel, there's an interaction. As shown in Figure 1-6, the factors interact little, if at all, with respect to

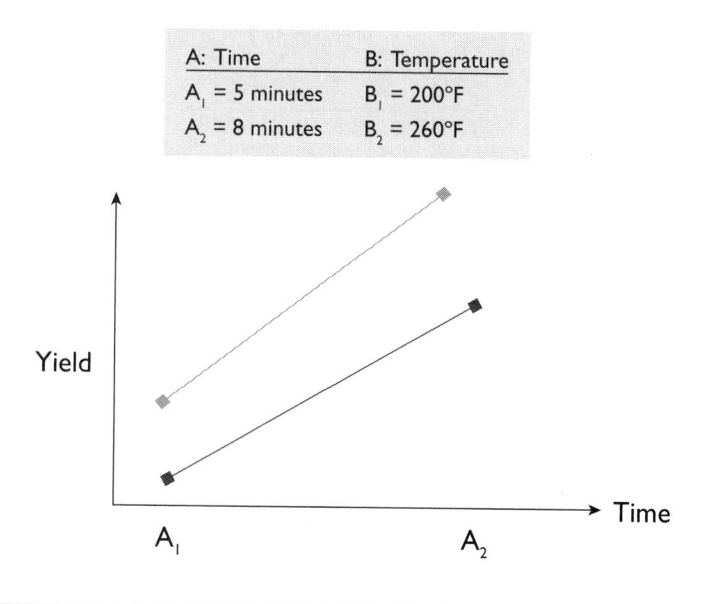

Figure 1-6 The interaction between color and temperature; the effect of cook time is to increase color at either temperature.

the color engineering metric. Similar results would follow if the CTQ were another engineering metric such as moisture or density. In this example, with an engineering metric as Critical-to-Quality rather than a customer metric as Voice-of-Customer, the size of interactions can be reduced.

In sum, the Voice-of-Customer is what the customer wants. Systems transform the intent into the perceived result for VOCs. The perceived result is what the customer gets. A Critical-to-Quality characteristic is an engineering metric that quantifies the output of the system. A VOC is expressed in nontechnical terms and is frequently subjective. CTQ characteristics are expressed in technical terms and should (1) be related to the perceived results for VOCs, (2) quantify energy transfer, and (3) be an engineering metric.

1.6 DETERMINE CONTROL AND NOISE FACTORS

To make products affordable, engineers need to determine how to control the CTQ characteristics at minimal cost. The fifth step in the robust design process is to develop a list of control and noise factors for each CTQ. This section covers the following:

- Definition of control factors
- Definition of noise factors
- Sources of noise

In robust design, engineering parameters related to CTQs are categorized as either control factors or noise factors (see Figure 1-7). The Engineered System, or P-Diagram (see Chapter 6), for a product or process is a diagram that shows the relationship among system (or subsystem) parts, the CTQ, and the control and noise factors.

Control Factors: Parameters whose nominal values can be cost-effectively adjusted by the engineer.

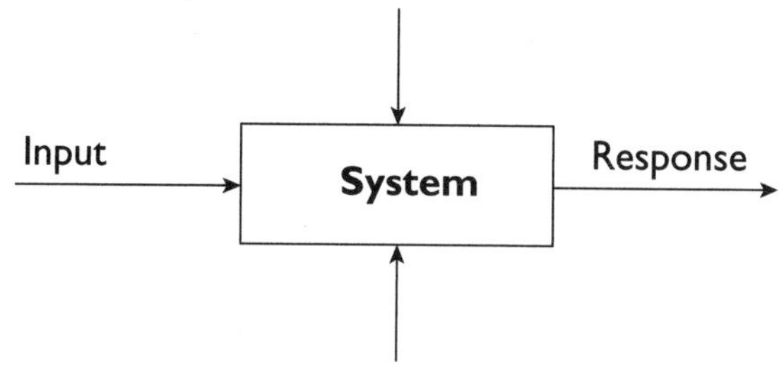

Noise Factors: Parameters that influence system variability but are difficult, expensive, or impossible to control.

Figure 1-7 Control and noise factors.

Brainstorming is a useful tool for developing an initial list of control and noise factors. Further investigation may be needed to research creative ideas that result or to discover additional factors. If the list of influential control and/or noise factors becomes prohibitively long, consider narrowing the scope of the study to a simpler subsystem. Then, you may need to redefine the response to establish a complete situational understanding of a wide range of data where several control factors may be interacting at once to produce an outcome.

Determining whether a factor is a noise or a control one often depends on the team's objective or the scope of the project. A factor considered control in some cases might be considered noise in others. For example, consider the material hardness factor (measured in Rockwell units). Design engineers focus on the product, so they may categorize material hardness as a control factor. However, process engineers focus on the process, so they may categorize material hardness as a noise factor; from their perspective, the process needs to be insensitive to the hardness of the material.

There are many sources of noise. Figure 1-8 shows five broad noise factor categories. Frequently, customer usage creates the most variability; but when developing a noise strategy, consider all possible sources to ensure that influential noise factors are not overlooked. Typically, control factors are obvious to the engineer because they relate directly to system design. On the other hand, it is easy to overlook some noise factors because they are often external to system design. Examining the five potential sources of noise can help engineers develop a thorough list of noise factors.

- Internal environment as a result of neighboring subsystems
- Part-to-part, or piece-to-piece, manufacturing variation
- Customer usage and duty cycle
- External environment
- Age, or deterioration

Figure 1-8 Source of noise factors.

In sum, control factors are parameters whose nominal values can be adjusted by the engineer, ideally with minimal impact on cost. A noise factor is a source of variability, either internal or external to the system. A noise factor disrupts the transfer of energy to the intended function.

1.7 Assign Control Factors to the Inner Array

The sixth step in the robust design process is to assign control factors to an inner array. Orthogonal arrays are efficient tools for multifactor experimentation. The orthogonal array of control factors is called the inner array. The *inner array* specifies the combinations of control factor levels to be tested. When running a designed experiment, the present design (or other reference design) should be included, allowing comparison of the current process to alternatives based on common testing conditions.

Figure 1-9 shows a sample orthogonal array. Genichi Taguchi (1992, 1999, 2000) made a significant contribution by adapting fractional factorial

Test	Control Factors							Response
Runs	A	B	C	D	E	F	F	Results
1	1	1	1	1	1	1	1	(R_1)
2	1	1	1	2	2	1	2	(R_2)
3	1	2	2	1	1	1	2	(R_3)
4	1	2	2	2	2	1	1	(R_4)
5	2	1	2	1	2	1	2	(R_5)
6	2	1	2	2	1	1	1	(R_6)
7	2	2	1	1	2	1	1	(R_7)
8	2	2	1	2	1	1	2	(R_8)

Figure 1-9 An orthogonal array.

orthogonal arrays (balanced both ways) to experimental design so that time and cost of experimentation are reduced while validity and reproducibility are maintained.

Taguchi's approach is disciplined and structured to make it easy for quality engineers to apply. Use of orthogonal arrays has been demonstrated to produce efficient robust designs that improve product development productivity. A full factorial design with seven factors at two levels would require $2^7 = 128$ experiments. Taguchi's L_8 orthogonal array requires only eight experiments. Typically, orthogonal arrays include a configuration in which all factors are set at Level 1. When using these arrays, the team may elect to let Level 1 represent the present design so that no testing beyond that specified in the inner array is necessary. Other teams may prefer to assign levels in increasing order of the factor settings so that it is easy to interpret the response tables and plots relative to the settings.

When testing at only two levels, the team may opt to test at levels above and below the present level. If this is preferred, the reference design should be run in addition to those specified in the inner array. Although the reference design statistics will be used for comparison to the selected optimal, they should not be included when developing response tables and plots.

Based on their relative impact on the system and on available resources, the team must now select the control factors for experimentation. They should then identify each factor's level and assign the factors and levels to the inner array. Previous experience, studies, or screening experiments can be used to help prioritize the brainstormed list of control factors.

Parameter Design experiments should be conducted with low-cost alternatives to present design settings (see Chapter 7). (Higher-cost alternatives are considered in Chapter 8, Tolerance Design, which emphasizes cost and quality tradeoffs.) As many factors as possible should be identified to enhance improvement potential. Control factors are usually tested at two or three levels in orthogonal array experiments; however, techniques are available to accommodate more levels.

The range of levels should be broad but still maintain system function. If the system ceases to function at a combination of factor levels designated by the inner array, data will be unavailable for a run. As a result, balance will be lost and all affect estimates will be biased.

In this book, control factor levels are denoted with numerals. Thus, for Level 2 factors, the levels are denoted 1 and 2 (or for Factor A, A_1 and A_2); and for Level 3 factors, the levels are denoted 1, 2, and 3 (or for A_1, A_2, and A_3).

EXAMPLE 1.2

Let's say you are an engineer working on the ball-swirling line at the Marion Bearing Manufacturing plant. As a cost-saving measure, management would like to loosen the ball bearing diameter's tolerance. As shown in Figure 1-10, the swirl-time variation has a significant impact on bearing quality.

- ➢ Short design life
- ➢ Poor damping, leading to high amplification factors
- ➢ Poor load-carrying capacity
- ➢ Limited maximum rotation speed

Figure 1-10 Quality problems caused by variations in swirl time.

The Six Sigma project champion has assigned your team the task of establishing process parameter nominal values at which the swirl time in the swirling machine's funnel will be robust to variation in ball bearing diameter in order to achieve a target in 10 seconds. After lab or project group assignments have been made, your team will be given an experimental apparatus with which to take data.

For each run, one group member will release the ball, another will time the ball with a stopwatch, and a third will record the result on paper. The experimental apparatus consists of a funnel mounted on a stand, a ramp mounted above the funnel, and a ball bearing. Your task in this lab is to conduct an experiment to determine how long it takes the ball bearing to roll down the funnel. Teams will perform one set of three runs with each possible pairing of group members. In other words, if there are four in your group, 12 sets of 3 runs each will be done.

A data sheet on which to record the observations is provided. For each run, you record the set number, the name of the persons releasing and timing, the order of the run in the set (1 to 3), and the time. Table 1-1 shows the control factors and levels that are most likely to impact energy transfer in the ball-swirling process. Levels are the different settings a factor can have. For example, if you want to determine how the response (swirl time) is affected by the

Table 1-1 Control Factors for Ball-Swirling Line

Control Factors	**Level 1**	**Level 2**
L: Run length	900mm	600mm
A: Ramp-to-funnel angle	30 degrees	45 degrees
H: Run end height	500mm	600mm
C: Clamping (unscrewed)	0.0 turns	0.5 turns
O: Operator training	Yes	No

Run No.	A	H	O	C	C5	C6	L
1	1	1	1	1	1	1	1
2	1	1	1	2	2	2	2
3	1	2	2	1	1	2	2
4	1	2	2	2	2	1	1
5	2	1	2	1	2	1	2
6	2	1	2	2	1	2	1
7	2	2	1	1	2	2	1
8	2	2	1	2	1	1	2

Figure 1-11 Use of an L_8 orthogonal array for a swirling machine's robust design.

factor (run length), you would need to set the factor at different levels (e.g., 900mm and 600mm).

For maximal test efficiency, this particular team elected to use the L_8 array for the inner array of their experimental plan. As shown in Figure 1-11, the ramp-to-funnel angle was assigned to column 1 to limit the number of changes necessary. It was believed that operator training would not interact with any of the other factors, so it was assigned to column 3, which then keeps the other four main effects free from confounding by any potentially real control-by-control interactions.

As Figure 1-11 shows, an L_8 orthogonal array enables selection of up to seven factors for testing with only eight runs. In comparison, a 2^K full factorial design of experiment (DOE)[1] requires 128 runs. A full factorial

1. A *design of experiment* is a structured, organized method for determining the relationship between design factors (*X*s) affecting a product and the output of that product (*Y*).

DOE measures the response of every possible combination of factors and factor levels. These responses are analyzed to provide information about every main effect and every interaction effect.

A full factorial DOE is practical when fewer than five factors are being investigated. Testing all combinations of factor levels becomes too expensive and time-consuming with five or more factors. Orthogonal arrays include selected combinations of factors and levels. It is a carefully prescribed and representative subset of a full factorial design. By reducing the number of runs, orthogonal arrays will not be able to evaluate the impact of some of the factors independently. In general, higher-order interactions are confounded with main effects or lower-order interactions. Because higher-order interactions are rare, usually the assumption is that their effect is minimal and that the observed effect is caused by the main effect or lower-level interaction.

If more than seven factors were selected for testing, it may have been more practical to use the L_{12} array, which is described in Chapter 2. As discussed there, use of L_{12}, L_{18}, L_{36}, or L_{54} orthogonal arrays is recommended. These arrays allow for testing many factors and share the quality that only fractions of interaction effects confound the main effects in any column.

1.8 SUMMARY AND ROAD MAP

The road map for engineering robust products with Six Sigma is shown in Figure 1-12. Chapters 2, 3, and 4 discuss in detail how to establish the Voice-of-Customer models and how to convert them into CTQs, design concepts, and design controls. CTQs represent the product or service characteristics that are defined by the customer (internal or external), which may include the upper- and lower-specification limits or any other factors related to them. A CTQ characteristic—what the customer

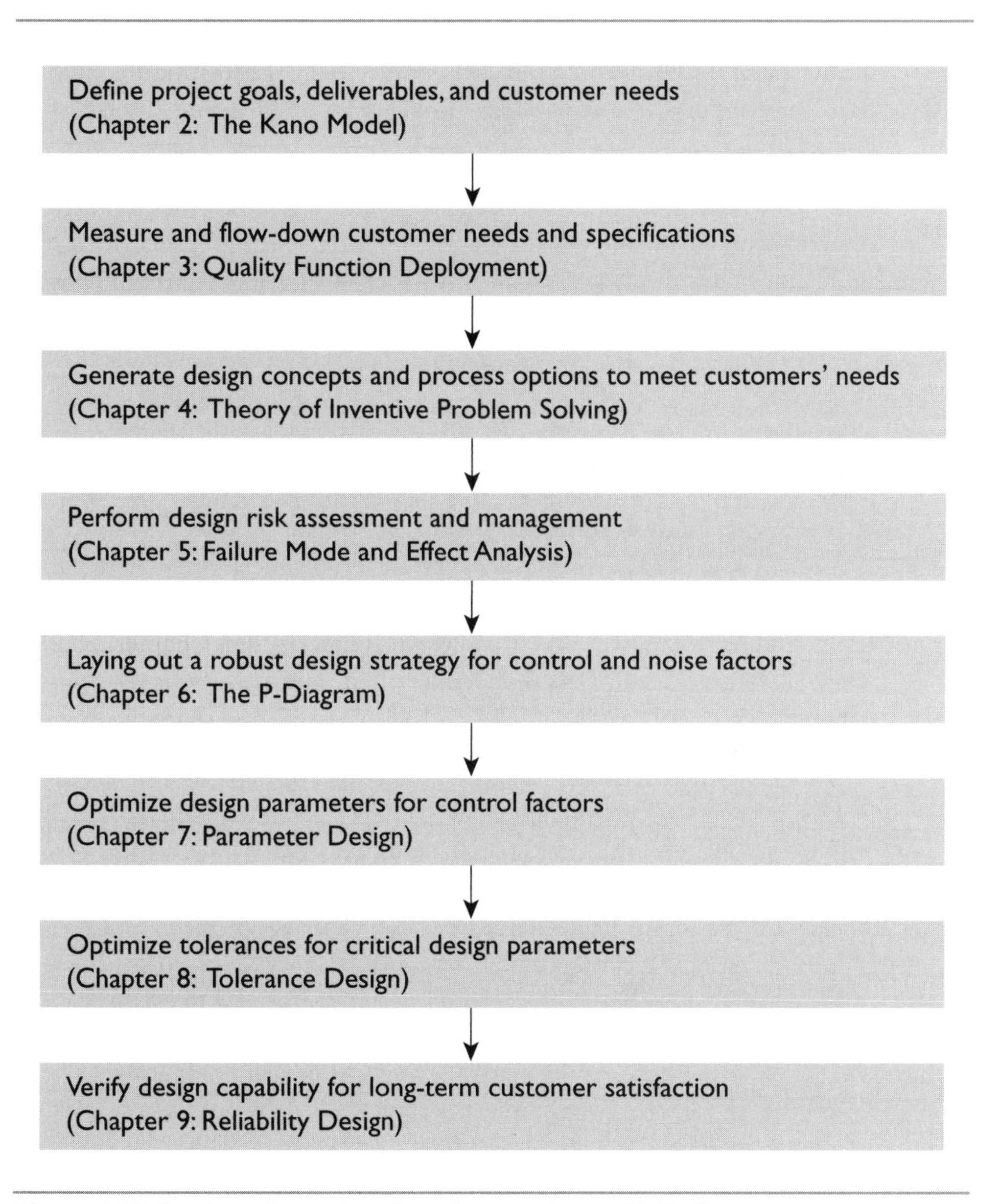

Figure 1-12 Road map for engineering robust products with Six Sigma.

expects of a product—usually must be translated from a qualitative customer statement into an actionable, quantitative business specification. It is up to engineers to convert CTQs into measurable terms using Six Sigma tools.

Six Sigma robust design starts with the voices of the customers, which reflect their spoken and unspoken needs and/or requirements. The Kano model, which is discussed in Chapter 2, helps engineers identify VOCs systematically.

Based on VOCs, a House of Quality can be built using a Six Sigma methodology called Quality Function Deployment (QFD). Within the House of Quality, customer requirements are converted into Critical-to-Quality characteristics. QFD enables identification and prioritization of the CTQs. A case study example in Chapter 3 illustrates the six steps to construct a House of Quality. The QFD process can also identify technical contradictions, which are the basis for applying the Theory of Inventive Problem Solving (TRIZ—the Russian acronym for the theory) to generate creative design concepts that can eliminate contradictions (see Chapter 4).

Critical-to-Quality characteristics reveal a main difficulty for developing robust designs. However, being able to integrate value-added features using TRIZ enables engineers to determine the final robust design concept. As illustrated with a practical example in Chapter 4, the quality of the design concept is the design's *DNA*, which drives product robustness.

Starting with Chapter 5, the focus shifts from control factors to noise factors, which are the process inputs that consistently cause variation in the output measurement that is random and expected and, therefore, not controlled. Strategies to manage noise (e.g., white noise, random variations, common-cause and special-cause variations, uncontrollable variables) are discussed in detail in Chapters 5, 6, 7, 8, and 9.

BIBLIOGRAPHY

Box, George E. P., and Norman R. Draper. *Empirical Model-Building and Response Surfaces.* New York: John Wiley & Sons, 1987.

Box, George E. P., William. G. Hunter, and J. Stuart Hunter. *Statistics for Experiments—An Introduction to Design, Data Analysis, and Model Building.* New York: John Wiley & Sons, 1978.

Cuthbert, Daniel, and Fred S. Wood. *Fitting Equations to Data.* New York: John Wiley & Sons, 1980.

Lipson, C. *Statistical Design and Analysis of Engineering Experiments.* New York: McGraw-Hill, 1973.

Lorenzen, Thomas, and Virgil Anderson. *Design of Experiments: A No-Name Approach.* New York: Marcel Dekker, 1993.

Montgomery, Douglas C. *Design and Analysis of Experiments, Fifth Edition.* New York: John Wiley & Sons, 2000.

Ray, Ranjit K. *Design of Experiments Using the Taguchi Approach.* New York: John Wiley & Sons, 2001.

Ross, Phillip J. *Taguchi Techniques for Quality Engineering.* New York: McGraw-Hill, 1995.

Taguchi, G. *The System of Experimental Design: Engineering Methods to Optimize Quality and Minimize Costs.* Dearborn, MI: Quality Resources, 1987.

Taguchi, G., S. Chowdhury, and S. Taguchi. *Robust Engineering: Learn How to Boost Quality While Reducing Costs and Time to Market.* New York: McGraw-Hill, 1999.

Taguchi, G., S. Chowdhury, and Y. Wu. *The Mahalanobis-Taguchi System.* New York: McGraw-Hill, 2000.

Taguchi, G., and S. Tsai. *Taguchi on Robust Technology Development.* New York: American Society of Mechanical Engineers, 1992.

Ueno, K. "Companywide Implementation of Robust Technology Development." Proceedings of the American Society of Mechanical Engineers, New York, March 1997.

2 The Kano Model

Listening to the Voice-of-Customers

The Six Sigma journey toward developing robust designs starts by listening to the Voice-of-Customers (VOCs), which reflect their spoken and unspoken needs or requirements. The VOCs can be heard in a variety of ways: direct discussion or interviews, surveys, focus groups, customer specifications, observation, warranty data, field reports, complaint logs, and so on. However, a major challenge exists for completely capturing the Voice-of-Customers; although some are spoken, some VOCs are actually unspoken. In addition, some customer requirements have nonlinear relationships with product performance, which presents the focus for robust design. This chapter describes the Kano Model, which helps identify the Voice-of-Customers systematically.

The notion of inherent quality of products and services, which are deemed to be superior as opposed to inferior, has been discussed and debated for centuries. Philosophers, such as Aristotle, Rene Descartes, and John Locke, have provided different facets for the definition of *quality.* In the 1930s, Dr. Walter A. Shewhart (1986) began developing his definition of quality through the use of statistics and what is now termed *statistical quality control.* During and after World War II, statistical variations on the meaning of quality continued in the United States and Japan with the work of W. E. Deming (1982), Joseph Juran (1998), and Armand V. Feigenbaum (1991). In Japan, the work of Kaoru Ishikawa (1985), Shigeru Mizuno (1988), Shoji Shiba (1993), Yoji Akao (1990), and Genechi Taguchi (1987) provided additional perspectives and a much larger context in which quality is germane—for example, Total Quality Management (TQM) and Loss to Society, which both have greatly influenced today's theories about customer satisfaction.

2.1 HOW TO MAKE THE CUSTOMER HAPPY

In the late 1970s, Dr. Noriaki Kano of Tokyo Rika University further refined the notion of quality that he derived partially from his study of Herzberg's Motivator–Hygiene Theory, the premise of which was that jobs have specific factors that are related to job satisfaction or dissatisfaction. Herzberg (1976) noted that the elements related to job content could be divided into two categories: hygiene factors and motivators. He found that hygiene factors, such as job security, salary, and company policies, were important in reducing job dissatisfaction but would not necessarily provide job satisfaction. To promote job satisfaction, the employer must use motivator factors such as opportunity for advancement, recognition, responsibility, achievement, and doing quality work. Herzberg's theory is specifically job-related and reflects some of the distinct things that people want, or miss, while they work. According to him, hygiene factors—the lack of which cause dissatisfaction—must be present in the job before motivators can be used to stimulate people.

Whereas many of the previous definitions of quality were linear and one dimensional in nature (e.g., good or bad, small-versus-large Loss to Society), Dr. Kano integrated quality along two dimensions. The first is the degree to which a product or service performs, and the second is the degree to which the user is satisfied.

As shown in Figure 2-1, the juxtaposing of the quality parameters of user satisfaction and performance in a two-axis plot created the ability to define quality in a more sophisticated and holistic manner. The Kano Model is very useful in providing a level of sophistication not available in a one-dimensional model of quality. If the level of customer satisfaction is plotted on a vertical axis and the degree that the product or service has achieved a given performance attribute is plotted on the horizontal axis,

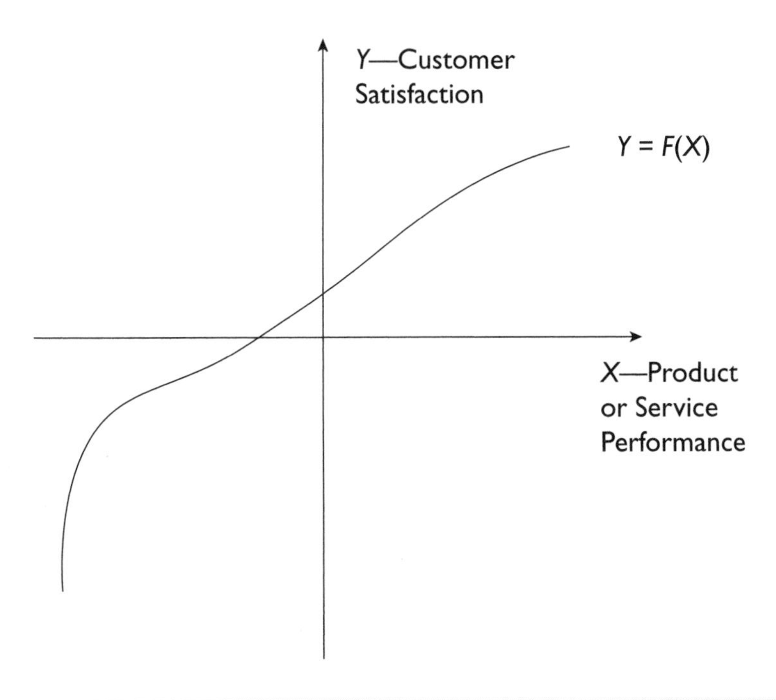

Figure 2-1 The Kano Model's two-axis plot of user satisfaction versus product performance.

different types of customer wants and needs can be shown to cause widely different responses.

The correlation of quality on two axes led Kano and associates (1984) to the following three unique definitions of quality:

Basic quality: Basic needs must be met in terms of customer satisfaction. If these needs go unfilled, customers will certainly be unhappy. Total absence or poor performance in any of the basic attributes could result in extreme dissatisfaction.

Performance quality: The performance attributes have a linear effect on customer satisfaction (i.e., more is generally better and good performance in these areas will improve customer satisfaction). Additionally, the price customers are willing to pay for goods and/or services is closely tied to these attributes.

Excitement quality: Excitement attributes are unexpected by the customer but, when present, can result in high levels of customer satisfaction.

The model shows that the customers' responses can be classified into three quality types, as noted in Figure 2-2.

2.1.1 Basic Quality

The dynamics of *basic quality* indicate that some customer requirements, if not achieved, result in a high level of dissatisfaction; and, if achieved, they have only a limited effect on customer satisfaction. The reason for this is that the customer expects this type of quality. For example, when going into a restaurant for a meal, the customer expects there to be a place setting; if there isn't one, the customer will be dissatisfied. If there is a place setting, no credit will be given because there is supposed to be

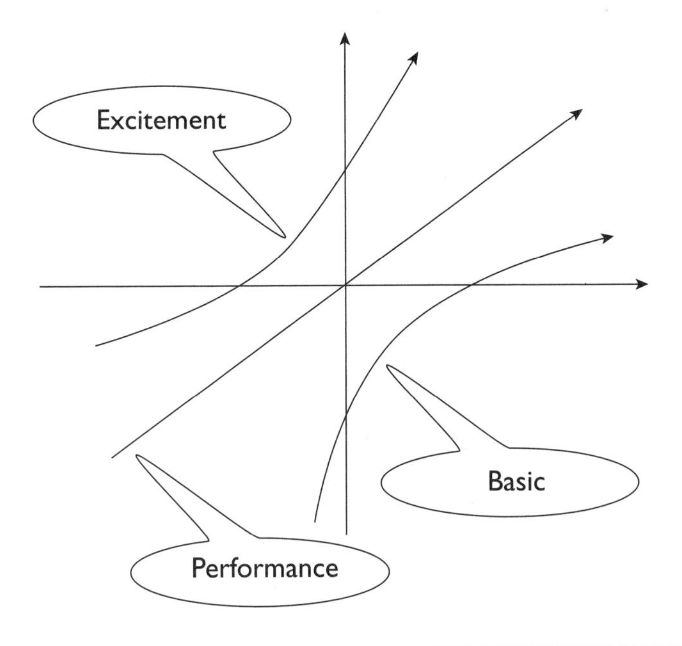

Figure 2-2 The Kano Model's types of quality.

one. On the other hand, having many place settings does not create any additional satisfaction.

In the automotive world, the customer expects a vehicle to start easily; to provide a safe driving environment; and to be free of squeaks, rattles, and excessive noise. Satisfaction is not necessarily created if a vehicle performs as expected. However, if *basic* needs are not met, the result could be devastating to the reputation and business of the original equipment manufacturer (OEM). Basic quality provides "down-side risk" with very little "up-side potential" for customer satisfaction.

Not meeting basic quality attributes is often interpreted by the customer as total failure and is expressed by significant complaining. In industry, achievement of basic quality is typically measured by customer complaints, warranty data, product recalls, number of lawsuits, and things-gone-wrong (TGW) and other failure reports.

2.1.2 PERFORMANCE QUALITY

The second type of customer requirement generates satisfaction in proportion to the performance of the product. This quality type is referred to as *performance quality,* the attributes of which generally cause a linear response. Increased levels of achievement cause increased levels of satisfaction.

The customer in a restaurant expects his or her order to be taken promptly and accurately and the food to be delivered in a reasonable period of time. The better the restaurant meets these needs, the more satisfied the customer is. The restaurant wait staff and management frequently ask patrons how the meal is while they are dining, and most customers freely express their opinions relative to performance quality when they are asked. This type of information is often called Voice-of-Customers data, because they are the types of things customers like to talk about.

Vehicle owners want the car to perform one way or another and to have this or that feature. Measurements are taken using customer research tools, feature rating surveys, and ride/drive evaluations that ask how well a product performs relative to a graduated scale. Although customers expect a vehicle's engine to run well, such performance is gauged relative to expectations. Someone who buys a small economy car will not expect the same raw performance as a person who buys a "muscle" car. Generally speaking, however, the better the performance, the greater the satisfaction.

2.1.3 Excitement Quality

The third quality type generates positive satisfaction at any level of execution, and it is referred to as *excitement quality.* Excitement is generated because customers receive some feature or attribute that they did not expect, ask for, or even think was possible. For example, if the restaurant provides an on-the-house glass of champagne, the customer will be pleasantly surprised. Likewise, the owner of a vehicle may not expect it to have a built-in GPS, a maintenance-free battery, heated seats, and so on, but he or she will be pleased when such features are discovered during ownership.

Customers generally do not articulate excitement attributes in customer surveys because they do not know that they want special features. To generate customer excitement and brand loyalty, companies must leverage their creative resources to identify ideas and innovations that will lead to customer excitement. Excitement quality becomes the special reason why some customers make specific companies their default choice over the competition and return to buy again and again. Differentiation among products is achieved by high-level execution of performance attributes combined with one or more customer excitement features. Such a combination provides the greatest opportunity for competitive advantage (Figure 2-3).

Excitement attributes cause an exponential response. Small improvements in the provision of special items can result in relatively large increases in customer satisfaction. Several exciting features may accumulate and generate sheer delight on the part of customers.

The Kano Model is useful for providing a two-dimensional model of quality. In actual application, requirements do not always fall neatly into one of the three categories. Very high levels of performance relative to expectations can act like excitement attributes and provide real reasons to

- ➢ Improve customer satisfaction
- ➢ Increase product sales
- ➢ Develop new markets
- ➢ Improve product features' pricing
- ➢ Advertise product features
- ➢ Develop new products
- ➢ Channel product distribution
- ➢ Enhance product service

Figure 2-3 The importance of excitement quality.

choose a particular product over its competitor. Likewise, an intended excitement feature executed poorly will cause dissatisfaction.

2.2 CUSTOMER REQUIREMENTS OVER TIME

It has been observed that customers' requirements change over time. Sources of excitement when they were first introduced tend to become expected as the market becomes familiar and saturated with them. In time, excitement quality becomes a performance item and, with the passage of time, quite possibly a basic requirement.

Automatic transmissions, which initially provided excitement because they made cars much easier to drive, are classified today as a basic quality item. For a time, people made comparisons because some designs performed better than others, but now owners demand that automatic transmissions perform flawlessly. Customers talk about them only if there is a problem. It is critical to understand the dynamic features of the performance and excitement attributes to maintain competitive advantage. Figure 2-4 shows the dynamic of time.

To be competitive, there is no doubt that products or services must flawlessly execute all three of the quality types. Meeting customers' basic quality needs provides the foundation for the elimination of dissatisfaction and complaints. Exceeding customers' performance expectations creates a competitive advantage, and innovations differentiating the products and the organizations create an excited customer.

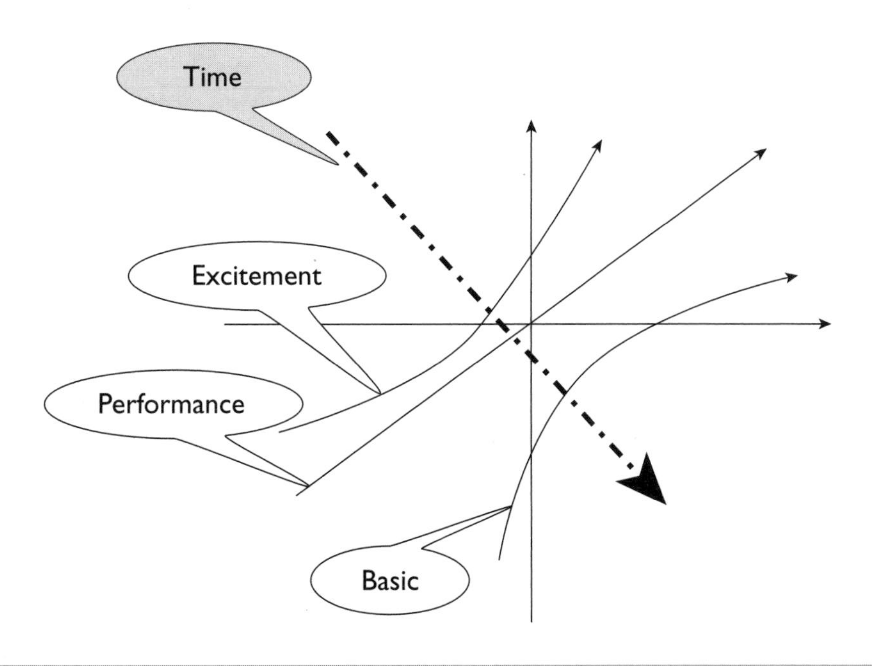

Figure 2-4 The Kano Model's quality types over time.

From customers' perspective, there are three metrics categories for meeting their requirements: the smaller-the-better, the nominal-the-best, and the larger-the-better. As noted in Chapter 1, each of these characteristics should be measurable on a continuous scale.

An important step in the Six Sigma robust design process is to identify control factor levels to ensure customer satisfaction. The next sections cover the following:

- Identifying control factor levels
- Analysis of noise-by-control interactions
- Reducing variability
- Assignment of noise factors to the outer array
- Selecting noise levels

2.3 CONTROL FACTOR LEVELS: ENSURE THREE TYPES OF QUALITY

To ensure basic quality, performance quality, and excitement quality, it is essential to identify control factor levels, which are the different settings a factor can have. For example, if engineers want to determine how the response (speed of data transmittal) is affected by the factor (connection type), they would need to set the factor at different levels (modem and local area network). Chapter 1 introduced the concept for orthogonal arrays, which are often employed in industrial experiments to study the effect of several control factors. Dr. Taguchi recommends using L_{12}, L_{18}, L_{36}, or L_{54} orthogonal arrays. Such arrays allow the testing of many factors and share the quality that only fractions of interaction effects confound the main effects in any column. By using these orthogonal arrays, it is possible to get satisfactory accuracy for robust design experiments cost-effectively.

Two-dimensional orthogonal arrays of numbers possess the interesting quality that, by choosing any two columns, you receive an even distribution of all the pairwise combinations of values in it. As shown in Figure 2-5, the L_{12} orthogonal array can accommodate up to 11 Level 2 factors. It is a very efficient orthogonal array for testing many factors and, as such, is especially useful in screening experiments.

A screening design of experiment (DOE), where L_8 and L_{12} orthogonal arrays are used, is a specific type of a fractional factorial experiment. Such a design—a low-resolution design—minimizes the number of runs required in an experiment. A screening DOE is practical when you can assume that all factors are known and are included, as appropriate, in the

Run No.	1	2	3	4	5	6	7	8	9	10	11
1	1	1	1	1	1	1	1	1	1	1	1
2	1	1	1	1	1	2	2	2	2	2	2
3	1	1	2	2	2	1	1	1	2	2	2
4	1	2	1	2	2	1	2	2	1	1	2
5	1	2	2	1	2	2	1	2	1	2	1
6	1	2	2	2	1	2	2	1	2	1	1
7	2	1	2	2	1	1	2	2	1	2	1
8	2	1	2	1	2	2	2	1	1	1	2
9	2	1	1	2	2	2	1	2	2	1	1
10	2	2	2	1	1	1	1	2	2	1	1
11	2	2	1	2	1	2	1	1	1	2	2
12	2	2	1	1	2	1	2	1	2	2	1

Figure 2-5 An L_{12} (2^{11}) orthogonal array.

experimental design. Screening DOEs are used to distinguish important design parameters (*X*s) from the many parameters that the thermal engineer has to consider, with a minimal number of experiments or simulations. An output of a screening DOE is a linear mathematical model that relates the response to the design parameters.

Because of the nature of the L_{12} array, columns cannot be combined to create a column with more than two levels. As with any orthogonal array, some columns can be left "empty" if less than the maximum number of factors is tested. If less than 11 factors are tested, allocate the factors to any column. Since fractions of the interaction effects are distributed across the columns, factors can be allocated to columns without regard to confounding.

Factor or interaction effects are said to be confounded when the effect of one factor is combined with that of another. In other words, the effects of multiple factors on a response cannot be separated. This occurs to some degree in all situations and least frequently when the data is obtained from a carefully planned and executed experiment having a predefined objective.

As shown in Figure 2-6, the L_{18} orthogonal array allows for testing up to seven Level 3 factors and one Level 2 factor. If less than seven Level 3 factors are desired, any Level 3 column can be modified using the "dummy" treatment. With the dummy treatment, replace Level 3 with either Level 1 or Level 2. When calculating the level averages, be sure to divide the corresponding level sum by 12 rather than 6.

In addition, if a factor with more than three levels is desired, columns 1 and 2 can be combined to produce a column that accommodates up to six levels. The Level 6 column replaces both columns 1 and 2. To create a Level 6 column, replace the six different combinations of levels in columns 1 and 2 with Levels 1 through 6. So, in the Level 6 column, Level 1

Run No.	1	2	3	4	5	6	7	8
1	1	1	1	1	1	1	1	1
2	1	1	2	2	2	2	2	2
3	1	1	3	3	3	3	3	3
4	1	2	1	1	2	2	3	3
5	1	2	2	2	3	3	1	1
6	1	2	3	3	1	1	2	2
7	1	3	1	2	1	3	2	3
8	1	3	2	3	2	1	3	1
9	1	3	3	1	3	2	1	2
10	2	1	1	3	3	2	2	2
11	2	1	2	1	1	3	3	2
12	2	1	3	2	2	1	1	3
13	2	2	1	2	3	1	3	2
14	2	2	2	3	1	2	1	3
15	2	2	3	1	2	3	2	1
16	2	3	1	3	2	3	1	2
17	2	3	2	1	3	1	2	3
18	2	3	3	2	1	2	3	1

Figure 2-6 An L_{18} ($2^1 \times 3^7$) orthogonal array.

will appear in rows 1 through 3, Level 2 in rows 4 through 6, and so forth, to Level 6 in rows 16 through 18.

In sum, the L_{18} accommodates up to seven Level 3 factors and one Level 2 factor. Fractions of interaction effects confound the main effects, which minimizes the impact of the confounding. This feature makes the L_{12} a really incredible array—11 factors in 12 runs with diminished impact of confounding. Control factors are allocated to the inner array, and noise factors are allocated to the outer array.

The inner array is sometimes referred to as the control array. The numbers inside the inner array represent the levels of the control factors. The symbols plus (+) and minus (–) are also used to denote factor levels. The rows of the inner array each represent an individual design configuration, or control factor combination, that will be tested.

2.4 NOISE-BY-CONTROL ANALYSIS: MAKING PRODUCTS MORE ROBUST

Fundamental to engineering robust products is the identification of control factors that interact with noise. Noise-by-control interaction analysis is one method for identifying a control factor level at which the system is more robust. When the effect of one factor depends on the level of another factor, the two factors interact. On an interaction plot, the less parallel the lines, the greater the interaction.

In Figure 2-7, an interaction plot compares the impact of one factor on the response at various levels of another factor. This plot depicts the interaction effect on Response y between Factors A and B. When B is at Level B–, the value of y is nearly the same at both levels of A. So, the effect of A when B is at B– is very small. When B is at Level B+, the difference in y between the A– and A+ levels is great. So, the effect of A when B is at B+ is larger than the effect of A when B is at B–. Therefore, Factors A and B interact, as evidenced by the nonparallel lines.

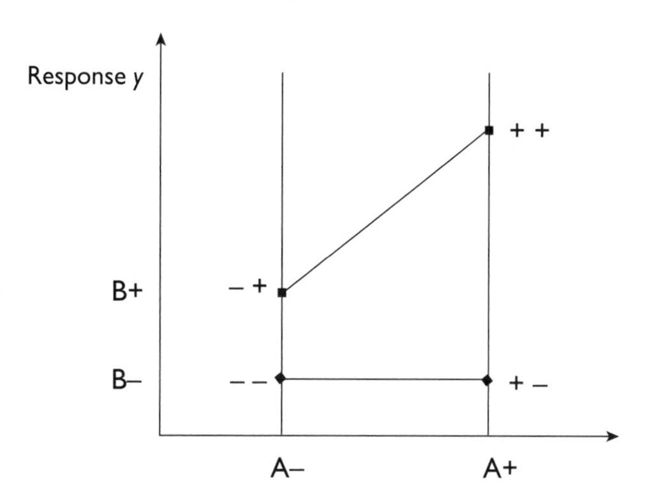

Figure 2-7 Intended function: transfer of energy to the intended result.

Example 2.1

Suppose optimized sweetness of a cup of tea is a critical indicator for performance quality. Here, the influence of the "sugar" and "stirring" factors on the sweetness of a cup of tea are investigated. Sugar is tested at the "no sugar" and "one cube" levels and stirring is tested at the "not stirred" and "stirred" levels. Figure 2-8 shows the interaction plot of the factor level combinations. The effect of sugar at one level of stirring is different from that at the other level—hence, the nonparallel lines. Since the effect of sugar depends on the level of stirring, note that these two factors interact with respect to sweetness—an important indicator of quality performance.

As shown in Figure 2-8, at the not stirred level of stirring, there is only a slight difference in sweetness between the no sugar and the one cube of sugar levels. Thus, the effect of sugar is small at the not stirred level of stirring. At the

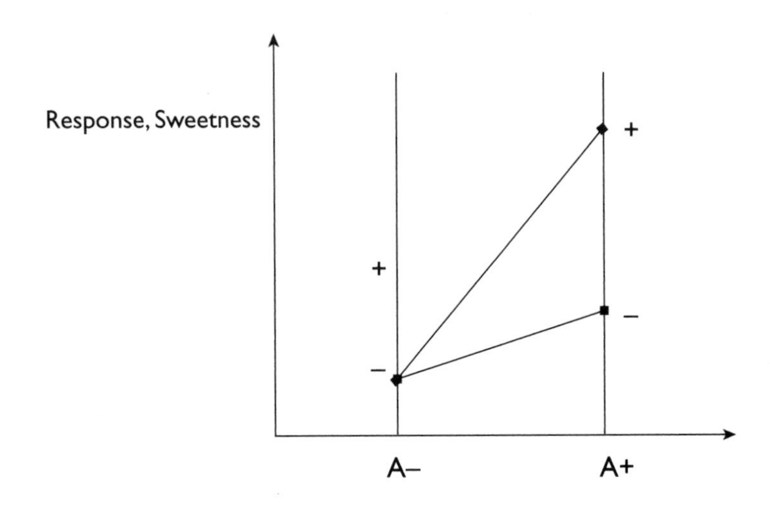

Figure 2-8 Influence of "sugar" and "stirring" on the sweetness of a cup of tea.

stirred level, the difference in sweetness between the two sugar levels is likely to be quite noticeable. Thus, the effect of sugar is large at the stirred level. Since the effect of sugar is different at the two stirring levels, note that the effect of sugar depends on the stirring level and that the factors interact.

2.5 VARIABILITY REDUCTION

Basic quality, performance quality, and excitement quality need to be ensured in the presence of various noise factors, which cause variation in product performance. According to the Kano Model, this variation will impact customer satisfaction.

In a robust design experiment, the configurations specified in the inner array are tested at the extremes of noise. Control factors that interact with the noise provide an opportunity for variability reduction. If a control factor interacts with noise, it may be possible to identify a control factor level at which the change in the response due to noise is minimal.

As shown in Figure 2-9, Factor C represents a control factor that was tested at levels C_1 and C_2. Factor N represents a noise factor; N_1 and N_2 are two extreme noise levels at which the system was tested. At C_2, the mean of the response distribution is closer to the target. But at this setting, because of the large variability caused by noise, many systems will still operate off-target. For this reason, the first priority is to set factor levels to reduce variation. Adjusting the mean to target is the second priority. This will be done with a different factor (like the one shown in Figure 2-10).

Consider the noise-by-control interaction plot shown in Figure 2-9. The nonparallel lines indicate an interaction between C and N. An interaction

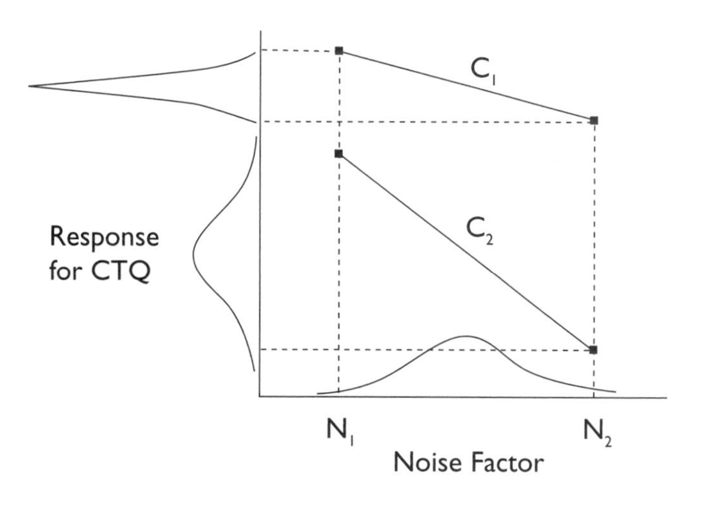

Figure 2-9 Variability reduction to ensure customer satisfaction.

occurs when the response achieved by one factor depends on the level of the other factor. On an interaction plot, when the lines are not parallel, there's an interaction. Because of this interaction, the effect of noise is less when Factor C is set at C_1 than at C_2. Thus, to reduce variability because of noise, N, set Factor C at Level 1.

The noise-by-control interaction plot of control Factor A and noise Factor N are depicted in Figure 2-10. Control Factor A does not interact with noise, but shifts the mean of the distribution. Because the variability in the response as a result of noise is the same regardless of the level of Factor A, this factor can be used to shift the mean toward the target without compromising robustness.

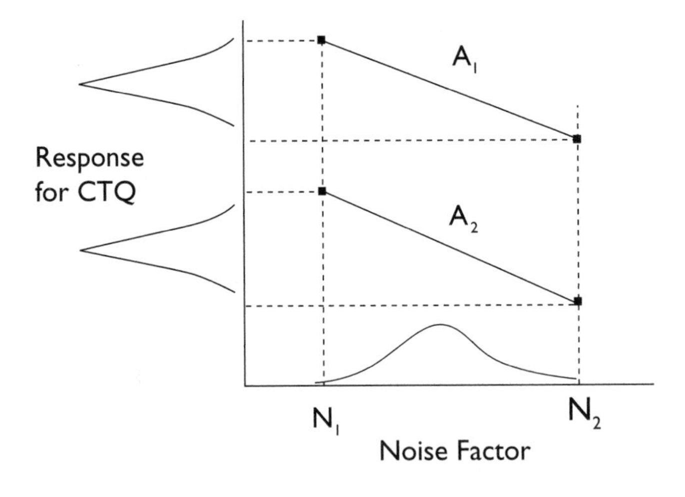

Figure 2-10 No reduction in variability.

2.6 THE OUTER ARRAY: AN ORTHOGONAL ARRAY SPECIFYING MULTIPLE NOISES

Planning fractional factorial experiments typically requires careful consideration of confounded effects on basic quality, performance quality, and/or excitement quality. If noise factors were included in the inner array, without proper allocation, the noise-by-control interactions could become confounded, making the robustness analysis less precise if not impossible. Therefore, for simplicity in designing the experiments, Dr. Taguchi recommends that noise factors be allocated to another array—the outer array.

In robust design methodology, repeated measurements of the response variable are often taken in a systematic fashion, with the goal to manipulate noise factors. The levels of those factors are then arranged in a so-called outer array; that is, an (orthogonal) experimental design. However, usually the repeated measurements are placed in separate columns (i.e., each is a different variable); thus, index i (in the formulas for the smaller-the-better, the larger-the-better, and the signed target) runs across the columns or variables of the data spreadsheet, or the levels of the factors in the outer array.

As shown in Figure 2-11, an outer array is an orthogonal array that specifies multiple noise (in dynamic designs, noise and signal) treatment conditions. Each combination of the control factors specified in the inner array is tested at every noise condition (in dynamic designs, noise and signal) specified in the outer array. By assigning noise factors to an outer orthogonal array, as recommended by Dr. Taguchi, you can saturate the inner array with control factors and keep all noise-by-control interactions free from confounding.

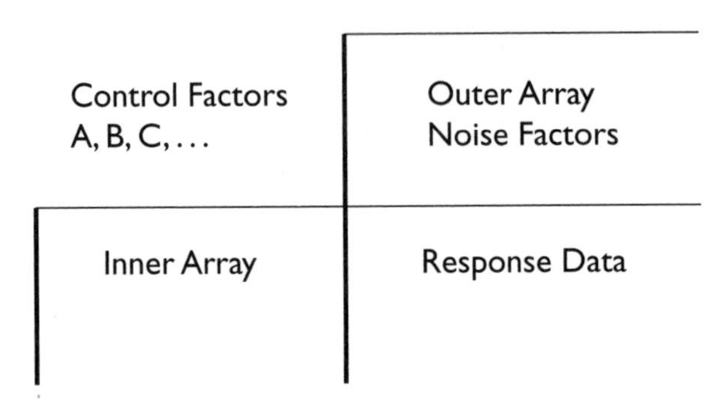

Figure 2-11 Differences between the inner array and the outer array.

EXAMPLE 2.2

Swirl time is an important performance indicator for the ball-swirling machine example in Chapter 1. The objective of the swirling machine experiment is to make swirl time *robust* to ball bearing diameter. Therefore, diameter, which is too difficult to control cost-effectively, is the only noise factor for this experiment.

At the present manufacturing capability, the ball bearings' diameters are expected to vary from 9mm to 14mm. Because these diameters represent the extreme noise conditions, the noise levels tested in the robust design experiment will be N_1 = 9mm and N_2 = 14mm. They make up the very simple outer array for this experiment (see Figure 2-12).

In sum, the outer array is used to specify the noise conditions of the experiment and keeps noise-by-control interaction effects free from confounding. The inner array specifies the control factor combinations for testing. If there are several strong noise factors, an efficient strategy for

Run No.	Inner Array					Outer Array	
	A	H	O	C	L	N_1: 9mm	N_2: 14mm
1	1	1	1	1	1		
2	1	1	1	2	2		
3	1	2	2	1	2		
4	1	2	2	2	1		
5	2	1	2	1	2		
6	2	1	2	2	1		
7	2	2	1	1	1		
8	2	2	1	2	2		

Figure 2-12 Assigning diameters as noise to the outer array.

testing noise must be developed. To this end, first identify the most influential noise factors and their levels.

2.7 Selecting Noise Levels

To satisfy Voice-of-Customers factors for basic quality, performance quality, and excitement quality, a Six Sigma project team needs to prioritize the list of brainstormed noise factors based on the following:

- The effect on energy transfer
- The effect on response variability
- Engineering knowledge
- Screening experiments
- Customer-usage profiles

Generally, manufacturing variability contributes less to system response variability than customer-usage conditions. Although noise factors are "uncontrollable" by definition, they must be mimicked during testing. Additionally, noise levels must be reproducible from run to run.

To establish the desired extremes of noise, it is usually sufficient to test each noise factor at two levels. As shown in Figure 2-13, it is crucial for noise levels to establish a noticeable separation in responses. Without noticeable separation, there is nothing in the experiment to which to become robust. On the other hand, noise levels should not be so extreme that they overwhelm the effects of the control factors. Screening experiments on the present design at various noise levels will provide evidence of noise separation.

If one noise factor dominates all of the others, it may be used as the surrogate for all. A surrogate noise factor must be such that if the system is robust to the surrogate, it is likely to be robust to other noise. A surrogate noise must (1) affect energy transfer and (2) create large response variability with respect to other noises.

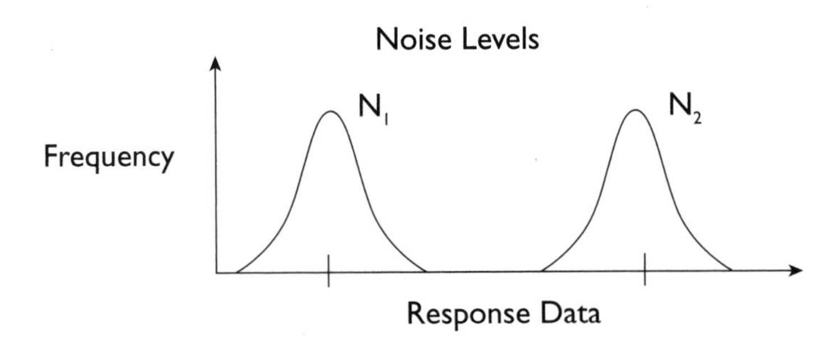

Figure 2-13 Noise levels establish a noticeable separation in responses.

Example 2.3

Computer Numerical Control (CNC) has been around since the early 1970s. The intent of CNC machining is to create shape and dimension by transforming energy to remove material. Variability in a part's material hardness will affect the energy transfer and cause variability in response. If the CNC machining process removes material consistently despite the presence of varying hardness, it is likely to do so in the presence of other noises, such as variability in lubricant. This means that material hardness can be used as the surrogate for other noises.

There are several important categories of noise. *Compounding* is a strategy by which noise levels are grouped to produce extreme noise combinations. Compounding noises is a powerful and efficient means of introducing several sources of noise into an experiment and is strongly recommended. It is important to group the factor levels that produce (1) a low response—call this group N_1—and (2) a high response—call this group N_2.

These grouped, or compounded, noises are treated in the outer array as a single noise with two extreme conditions. Without an effective noise strategy, achieving robustness is impossible. If there is uncertainty about the impact of noise factors, either individually or in combination, conduct a designed experiment with noise factors.

2.8 Summary

The Kano Model is a quality measurement tool that can be used to prioritize CTQ requirements based on their impact on customer satisfaction. The model defines customer satisfaction by dividing the outputs and/or service or product features into three groups: (1) basic requirements (the minimum a customer expects), (2) satisfiers (additional performance features that please customers), and (3) delighters (outputs or features that customers didn't expect that excite them).

Which identified requirements and control factors are important to customers can be determined by using the Kano Model. Such an analysis can help a Six Sigma team rank requirements for different customers from the highest to the lowest priority. The results can be used to prioritize the control factors to improve a company's chances of satisfying each customer's most important type of quality.

BIBLIOGRAPHY

Akao, Y. *Quality Function Deployment: Integrating Customer Requirements into Product Design.* Shelton, CT: Productivity Press, 1990.

Berger, C., R. Blauth, D. Boger, C. Bolster, G. Burchill, W. DuMochell, et al. "Kano's Methods for Understanding Customer-Defined Quality." *Center for Quality Management Journal,* 4: 3–36, 1993.

Deming, W. E. *Quality Productivity and Competitive Position.* Cambridge: Massachusetts Institute of Technology, 1982.

Feigenbaum, A. V. *Total Quality Control.* New York: McGraw-Hill, 1991.

Hauser, J. R., and D. Clausing. "The House of Quality." *Harvard Business Review,* May-June: 63–73, 1988.

Herzberg, F. *The Managerial Choice: To Be Efficient and to Be Human.* Homewood, IL: Dow Jones-Irwin, 1976.

Ishikawa, K. *What Is Total Quality Control?: The Japanese Way.* Upper Saddle River, NJ: Prentice Hall, 1985.

Jacobs, Randy. "Evaluating Satisfaction with Media Products and Services: An Attribute-Based Approach." *European Media Management Review,* Winter: 1–9, 1999.

Juran, J. M., and A. B. Godfrey. *Juran's Quality Handbook.* New York: McGraw-Hill, 1998.

Kano, N., N. Seraku, F. Takahashi, and S. Tsuji. "Attractive Quality and Must-Be Quality." *The Journal of the Japanese Society for Quality Control—Hinshitsu,* April: 39-48, 1984.

Mizuno, S. *Management for Quality Improvement: The Seven New QC Tools (Productivity's Shopfloor).* Shelton, CT: Productivity Press, 1988.

Pande, Peter, Robert Neuman, and Roland Cavanagh. *The Six Sigma Way Team Fieldbook: An Implementation Guide for Process Improvement Teams.* New York: McGraw-Hill, 2002.

Reichheld, F. F., and W. E. Sasser. "Zero-Defections: Quality Comes to Services." *Harvard Business Review,* September/October: S105–111, 1990.

Shiba, S., A. Graham, and D. Walden. *A New American TQM: Four Practical Revolutions in Management.* Portland, OR: Productivity Press, 1993.

Shewhart, W. *Statistical Method from the Viewpoint of Quality Control.* Mineola, NY: Dover Publications, 1986.

Taguchi, G. *The System of Experimental Design: Engineering Methods to Optimize Quality and Minimize Costs.* Dearborn, MI: Quality Resources, 1987.

3 Quality Function Deployment

Building a House of Quality

Based on the Voice-of-Customers (VOCs), a House of Quality can be built using a Six Sigma methodology called Quality Function Deployment (QFD). Within the House of Quality, customer requirements can be converted into Critical-to-Quality characteristics (CTQs), which are the key measurable characteristics of a product, process, or service. The performance standards or specification limits of the CTQs must be met to satisfy customers, and QFD facilitates identification and prioritization of them. Also, through QFD, it is possible to identify potential technical contradictions among CTQs. Technical contradictions are the basis for applying the Theory of Inventive Problem Solving (TRIZ) to generate creative design concepts (see Chapter 4).

Quality Function Deployment is a structured methodology and mathematical tool that is used to identify and quantify customers' requirements and to translate them into key critical parameters. In Six Sigma, QFD helps project teams prioritize actions to improve a process or product to meet customers' expectations. This chapter describes the six steps to use to construct a House of Quality and includes a case study example.

3.1 MARKET RESEARCH AND DETERMINING CUSTOMERS' NEEDS

As a systematic process, QFD motivates a Six Sigma project team to focus on its customers. Cross-functional teams use this technique to identify and to resolve issues involved in providing products, processes, services, and strategies. A prerequisite to Quality Function Deployment is *market research*—the process of understanding what the customer wants, how important its benefits are, and how well different providers of products that address the benefits are perceived to perform (see Figure 3-1). Market research is essential to QFD because it is impossible to provide products that will consistently attract customers unless you have a good understanding of what they want.

To understand customers' needs, a Six Sigma team should proceed as described in the sections that follow.

- Market studies and reports
- Interviews of current customers
- Lost-customer interviews
- Focus groups
- Surveys (e.g., in person, by telephone or mail, online)
- Customer feedback
- Complaint logs
- Observation of customers

Figure 3-1 Market research—understand what customers want.

3.1.1 PLAN THE COLLECTION OF CUSTOMER INFORMATION

Which sources will be used? Consider using customer requirements' documents, requests for proposals, requests for quotations, contracts, customer specification documents, customer meetings and/or interviews, focus groups and/or product clinics, user groups, surveys, observations, suggestions, and feedback from the field. It is good to gather data from both current and potential customers. Pay particular attention to leading customers because they are a better indicator of future needs.

Plan who will perform the data-collection activities. Prepare a schedule and make arrangements for where they will take place well in advance of any meetings, focus groups, surveys, and so on.

3.1.2 PREPARE FOR COLLECTION OF CUSTOMERS' NEEDS DATA

Identify what kind of information is required. Prepare agendas, question lists, survey forms, and focus group and/or user meeting presentations. If you anticipate collecting a significant amount of data from customers, plan for how it is to be stored and analyzed and in what form you will want reports generated. Involve someone from the information technology group to help with the steps in planning and preparing for data collection.

3.1.3 USING DATA-COLLECTION MECHANISMS

Determine customers' needs or requirements by using the mechanisms in the preceding steps. Document their needs. Consider tape recording customer meetings or focus groups. Be sure to ask "why" so that you can understand specific needs and can determine their "root" needs. Evaluate spoken needs and be on the alert for unspoken ones, which should be identified. Needs that customers assume you know, and therefore are not verbalized, can be identified through preparation of a goal tree (see Section 3.2).

Extract statements of needs from relevant documents. Summarize surveys and other data. Use techniques such as ranking, rating, paired comparisons, or conjoint analysis to determine the importance of customers' needs. Remember to gather customer information from any other sources a company has available for review.

3.1.4 ORGANIZE AND SUMMARIZE INFORMATION

Consolidate similar needs and restate as needed and organize them into categories. Break down general needs into more specific ones by probing for exactly what is needed. Maintain a dictionary of original meanings to avoid misinterpretation. Use function analysis to identify key unspoken but expected needs. This is where having included the information technology group in planning and preparing for data collection will pay off. Some sort of automated system will allow for efficient analysis, evaluation, and organization of data. Knowing how you want to analyze the information before you collect it makes for the most cost-effective and efficient process.

Once needs are summarized, consider whether to get further customer feedback with regard to priorities. Consider having additional meetings and focus groups, doing other surveys, and so on to determine such priorities and state them using a 1 to 5 rating scale. It may be helpful to use ranking techniques to develop priorities.

It is important to remember that there is no one monolithic Voice-of-Customer. Customers' voices are diverse. In consumer markets, there are bound to be a variety of needs. Even within one buying unit, there may be multiple voices (e.g., children versus parents). This applies to industrial and government markets as well. There may even be a multitude of customer voices within a single organization—the voice of procuring, the voice of users, the voice of support or maintenance groups. Diverse voices must be considered, reconciled, and balanced to develop a truly successful product.

3.2 Goal Tree: Spelling Out Unspoken Needs

Quality Function Deployment is a very important tool to help improve market share by reducing the gap between customers' desires and products' performance. The fundamental principle of QFD is to drive the design of a product or service by gathering all relevant customer information about their needs and wants through surveys, interviews, tests, benchmarks, and so on. This means that the primary function of QFD is to identify the most important product issues and parameters, to link priorities, and to target value back to customers.

A design team's goal is to target value while developing a robust product. A *goal* can also be defined as a customer voice—WHAT the customer is asking for or specifying. As long as it is known *how* to satisfy customers' wishes (WHATs) through the properties, parameters, and attributes of the products (HOWs), this primary function is being fulfilled. To establish the correlation between WHATs and HOWs in the QFD, the functional structure of products must be decomposed into its basic components. In every case, each one of the identified relevant WHATs should be supported by at least one of the basic components of the functional structure of already existing products.

To achieve the design team's objective, the goal tree structure is a very useful tool. A *goal tree* breaks down or stratifies ideas in progressively greater detail. The objective is to partition a big idea, or problem, into its smaller components, making the idea easier to understand or the problem easier to solve.

The primary or global useful function of an engineering system can be decomposed into subfunctions that are arranged as a hierarchy. In this case, the term *function* is defined as the input/output relationship in a technical system that fulfills a task. *Subfunctions* are therefore also input/output relationships that fulfill subtasks in a technical system. In a goal tree, functions are defined in terms of actions performed on objects. Here, the actions are described by verbs while the objects are described by parameters or things. The following are a few examples:

- To increase torque
- To transfer load
- To decrease rotational speed
- To cut metal
- To maintain a reliable power supply
- To prevent leakage

As shown in Figure 3-2, a goal tree should be developed to such an extent that each subfunction can be stated as an action on a functional parameter. Also, it is important to determine at least one HOW that is related to each and every one of the WHATs identified in the first stage of the QFD process. This is a nontrivial task and requires experienced designers who are able to identify those correlations. The goal tree for the system in Figure 3-2 was developed by breaking down the overall system goal into a set of necessary and sufficient subgoals. You need to continue the task of breaking down each subgoal until basic engineering specifications (HOWs) are established.

For instance, an engineer devising a pump must first control its output (in liters per second)—that's the pump's primary job (see Figure 3-2). Then he must control its temperature or it will overheat and fail—that's a crucial but secondary consideration. Then one needs to think about

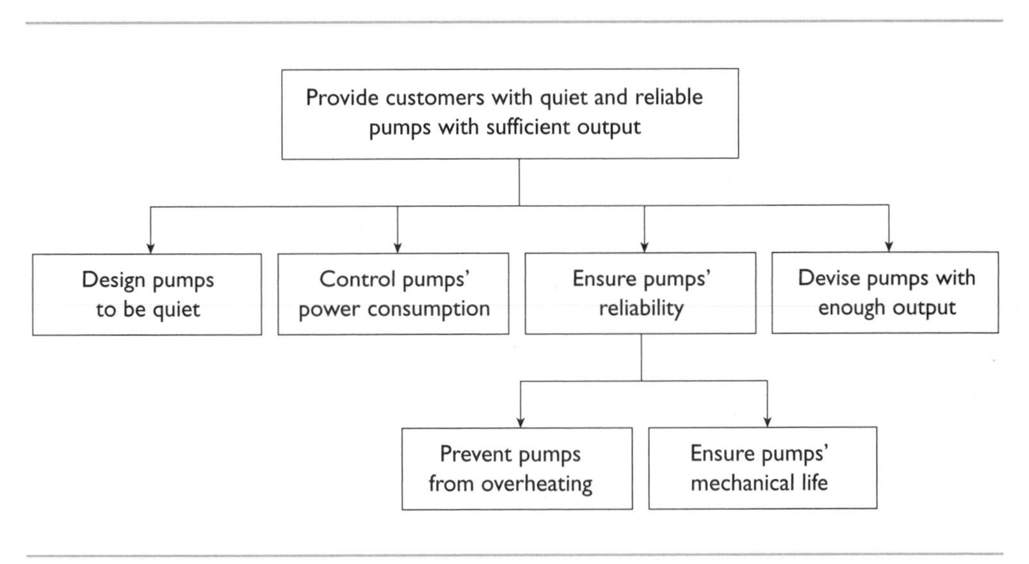

Figure 3-2 A sample goal tree for designing pumps.

controlling its noise, power consumption, reliability, and output. The third-order considerations to address are preventing overheating and ensuring their lifetime. For the pump's manufacturers, these are the qualities (along with price) that determine whether it will be a success in the marketplace. The design engineer will justifiably argue that she can't design a pump to be "quiet"—she needs to know how many decibels are allowed so that noise can be tested.

As illustrated in Figure 3-2, translating unmeasurable market goals into precise system specifications is part of the work of requirements analysis. In this case, it is easy to see why goals are thought to be vague, but the vagueness is only from the point of view of the system solution. From the perspective of the person purchasing or using the pump, quietness and reliability qualities are not vague at all.

It is essential to establish relationships between the WHATs and the HOWs of the already existing products through a goal tree analysis. After this, engineers can develop the functional parameters, or HOWs, of new products as part of the QFD process for building a House of Quality.

3.3 BUILDING A HOUSE OF QUALITY

Nearly every engineer would agree that customers are the most important part of a product—no customers, no need for product development. To be successful, design teams must know who the customers are and what the expectations are for the product being developed. When engineering robust products with Six Sigma, that process is known as identifying the Voice-of-Customers. The key to success in that process is gathering customer requirements' data and converting it into measurable critical-to-satisfaction elements.

Based on market research and goal tree analysis, QFD takes the VOC data and uses it to drive aspects of product development. Cross-functional Six Sigma teams develop matrices that analyze datasets according to the objectives of the QFD process.

Having captured VOC information in a variety of ways—direct discussions or interviews, surveys, focus groups, customer specifications, observations, warranty data, field reports, and so on—this understanding needs to be summarized in a product planning matrix, or House of Quality. The matrices are used to translate higher-level WHATs, or needs, into lower-level HOWs—technical responses to satisfy needs.

As shown in Figure 3-3, the House of Quality is the major matrix in a Quality Function Deployment process. When completed, a correlation matrix resembles a house structure and is often referred to as a "house" because its top is roof-shaped and sits above the main body. Such a matrix displays how the defined technical responses optimize or suboptimize each other.

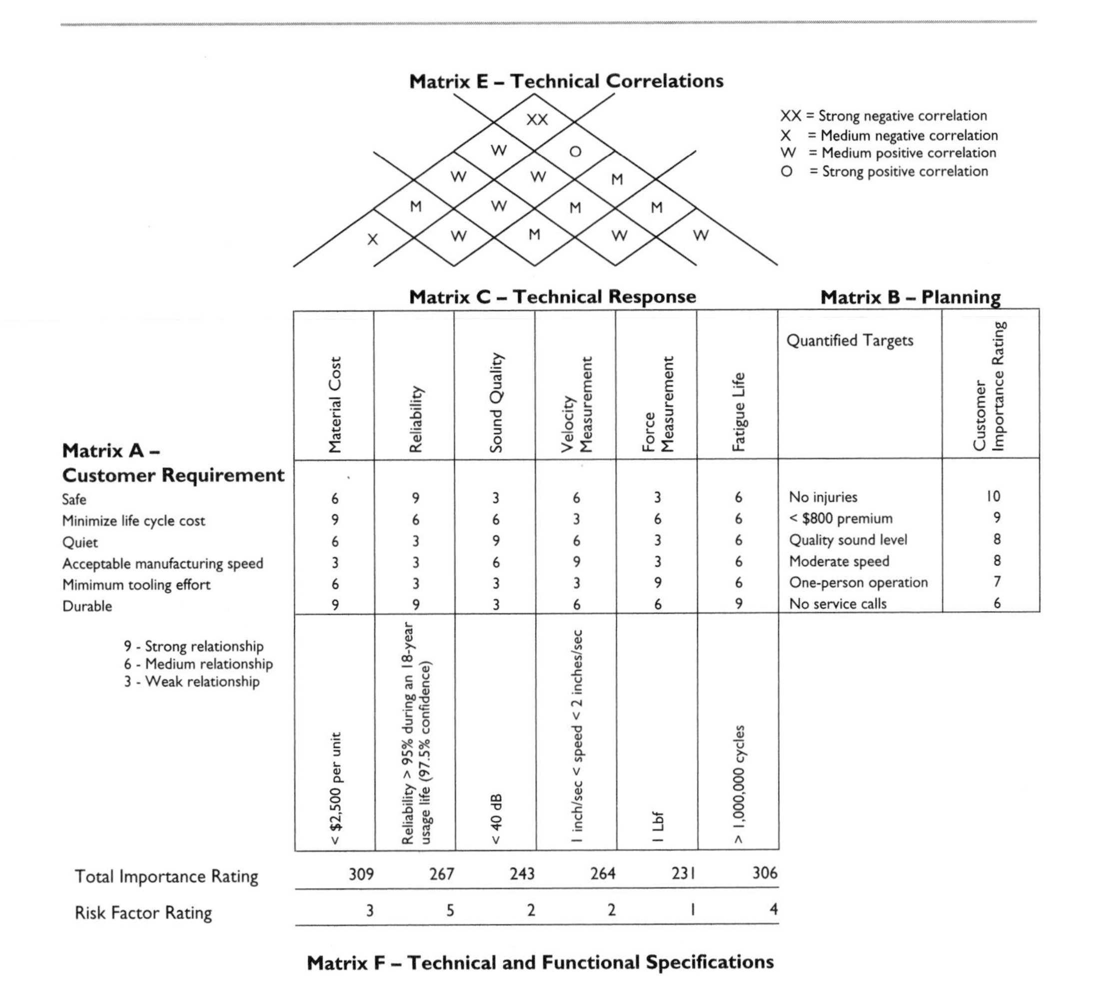

Matrix E – Technical Correlations
XX
W
O
W
W
M
M
W
M
M
X
W
M
W
W
XX = Strong negative correlation
X = Medium negative correlation
W = Medium positive correlation
O = Strong positive correlation
Matrix C – Technical Response
Matrix B – Planning
Matrix A – Customer Requirement
Material Cost
Reliability
Sound Quality
Velocity Measurement
Force Measurement
Fatigue Life
Quantified Targets
Customer Importance Rating
Safe 6 9 3 6 3 6 No injuries 10
Minimize life cycle cost 9 6 6 3 6 6 < $800 premium 9
Quiet 6 3 9 6 3 6 Quality sound level 8
Acceptable manufacturing speed 3 3 6 9 3 6 Moderate speed 8
Mimimum tooling effort 6 3 3 3 9 6 One-person operation 7
Durable 9 9 3 6 6 9 No service calls 6
9 - Strong relationship
6 - Medium relationship
3 - Weak relationship
< $2,500 per unit
Reliability > 95% during an 18-year usage life (97.5% confidence)
< 40 dB
1 inch/sec < speed < 2 inches/sec
1 Lbf
> 1,000,000 cycles
Total Importance Rating 309 267 243 264 231 306
Risk Factor Rating 3 5 2 2 1 4
Matrix F – Technical and Functional Specifications

Figure 3-3 A House of Quality for a ball bearing swirling machine.

Each house matrix is divided into several rooms. Typically you have customer requirements, design considerations, and design alternatives in a three-dimensional matrix to which weighted scores can be assigned based on the market research information collected. A House of Quality for a ball bearing swirling machine is illustrated in Figure 3-3. The sequence for building a House of Quality follows.

Step 1: Customer requirements are stated on the left side of the matrix, as shown in the figure. Ensure that the customer requirements included reflect the desired market segment(s) and address unspoken needs (assumed and excitement capabilities). If the number of customer requirements exceeds 20 to 30 items, decompose the matrix into smaller modules or subsystems to reduce the number of requirements in the matrix. For each need or requirement, state the importance of each using a 1 to 10 rating. Use ranking techniques to develop priorities.

Step 2: Next, establish technical responses to address customer requirements and organize them into related categories. The responses should be meaningful, measurable, and global. Technical responses should be stated in a way that avoids implying a particular technical solution so as not to constrain designers.

Step 3: Here you need to develop relationships between customer requirements and technical responses; use 9 for strong relationship, 6 for moderate relationship, and 3 for weak relationship. Be careful with the strong relationships because they often drive the Critical-to-Quality characteristics. Answer the questions: Have all customer needs or requirements been addressed? Are there any technical responses stated that don't relate to customer needs?

Step 4: Now is the time to do a technical evaluation of prior generations and others' products. Get access to competitors' products and perform product or technical benchmarking. Such an evaluation should be based on the technical responses that have been defined. Obtain other relevant data, such as warranty or service repair occurrences and costs, and consider this information in the technical evaluation. Proceed to developing preliminary functional specifications for technical responses.

Step 5: Use symbols for strong or weak, positive or negative relationships to show potential positive and negative interactions between technical responses. Too many positive interactions suggest potential redundancy in the *critical few* technical characteristics. Focus on negative interactions—consider product concepts or technology to overcome potential tradeoffs or consider the tradeoffs when establishing target values.

Step 6: Calculate the total importance ratings by quantifying them; assign weighting factors (9-6-3) to the strong, moderate, and weak relationships. Multiply the requirement's importance rating by the weighting factor in each box of the matrix and add the resulting products in each column.

Step 7: It is important to factor risk into the equation. Quantify risk by developing a difficulty rating (a 1- to 5-point scale, 5 being very difficult and risky) for each product requirement or technical characteristic. Consider technology maturity, personnel technical qualifications, business risk, manufacturing capability, supplier and/or subcontractor capability, cost, and schedule. Avoid too many high-risk items because this is likely to delay development and to exceed budgets. Be sure to assess the risk of whether difficult items can be accomplished within the project budget and schedule. Consider developing a risk-management plan to monitor complications associated with a product's development.

Step 8: Last, analyze the matrix and finalize the development strategy and product plans, including determination of required actions, areas of focus, and final functional specifications. determine answers to questions such as the following:

- Are the functional specifications' values properly set to reflect appropriate tradeoffs?
- Do functional specifications need to be adjusted with regard to the difficulty ratings and associated risks?
- Can risks be mitigated or reduced?
- Are they realistic with respect to price points, available technology, and difficulty ratings?
- Are the specifications reasonable with respect to the requirements' importance ratings?

It is also important to determine items for further Quality Function Deployment. To maintain focus on the critical few, less significant items can be ignored in subsequent QFD matrices.

One of the guidelines for successful QFD matrices is to keep the amount of information in each matrix at a manageable level. With a more complex product, if 100 potential needs or requirements were identified and were translated into an equal or even greater number of product requirements or technical characteristics, there would be more than 10,000 potential relationships to plan for and organize. That many relationships would be an impossible number to comprehend and to manage. It is suggested that individual matrices not address more than 20 or 30 items on each dimension.

Therefore, a larger, more complex product should have its customers' needs decomposed into component parts and processes to facilitate understanding of fundamental requirements (see Figure 3-4). The CTQ flow-down process places intense focus on the processes that impact customers. Six Sigma does this by developing the CTQ flow-down, which translates Voice-of-Customers into the Voice-of-Product Development and the Voice-of-Manufacturing Process.

Product QFD	Part QFD	Process QFD	Control Plan QFD
Define and prioritize customer needs	Identify critical parts and assemblies	Determine critical manufacturing processes	Review part and process CTQs
Analyze competitive opportunities	Flow-down product CTQs	Develop production equipment requirements	Establish process-control methods and parameters
Plan a product according to needs and opportunities	Translate into critical part and assembly CTQ target values	Establish manufacturing process CTQs	Establish inspection and test methods and parameters
Establish product CTQ target values			

Figure 3-4 Decomposition of the product into fundamental requirements.

When the operation of the product or the achievement of a performance characteristic can be mathematically related to a product or process design parameter, optimum product and process design parameters can be calculated. When these relationships are unknown, design of experiments (DOEs) can aid in determining the optimum parameter values and, thereby, developing a more robust design.

3.4 DEPLOY QUALITY THROUGH DESIGN OF EXPERIMENTS

A robust product is one that works as intended regardless of variation in a product's manufacturing process, variation resulting from deterioration, and/or variation in use. Robust designs can be achieved when the designer understands this and takes steps to desensitize the product to potential sources of variation. Robust design can be achieved through *brute-force* techniques—added design margin or tighter tolerances. Alternatively, robustness is possible through the use of *intelligent design,* which depends on understanding which product and process design parameters are critical to the achievement of a performance characteristic. It is important to determine what the optimum values are in order to achieve the performance characteristic and also to minimize variation. Such intelligent designs can be accomplished by building a House of Quality with the help of DOE techniques.

The design of experiments process is based on the objective of desensitizing a product's performance characteristic(s) to variation in critical product and process design parameters. Genichi Taguchi (1987) developed the concept of Loss to Society. In this concept, variability in critical design parameters is likely to increase the Loss to Society; and it is an expanded view of the traditional, internally oriented cost of quality. This is a quadratic relationship of increasing costs (Loss to Society) because the critical design values vary from the parameters' desired mean value.

To consider quality implications during design, the design process can be segmented into three stages. The first stage—System Design—establishes the functionality of the product, the physical product envelope, and general specifications. The second stage—Parameter Design—establishes specific values for design parameters related to physical and functional specifications. It is during the first two stages that the designer has the greatest opportunity to reduce product costs through effective functional design and parameter specification. The third stage—Tolerance Design—establishes acceptable tolerances around each parameter or target. Typically, this stage adds cost to the product through efforts to ensure compliance with the tolerances associated with product parameters.

Because an organization cannot cost-effectively instill quality into the product, engineers must focus on minimizing product variability through product and process design and control of processes. However, some variability is very difficult to control or even uncontrollable. As described in Chapter 1, the difficulty in controlling variation is referred to as noise. *Noise* is the result of variation in materials, processes, the environment, and products' use or misuse. Products need to be designed so that they are robust; that is, their performance is insensitive to naturally occurring, difficult-to-control variation—noise.

DOE techniques also provide a way to efficiently design industrial experiments that will improve understanding of the relationship between product and process parameters and the desired performance characteristic. When the design of the experiment is based on a fractional factorial process, it allows an experiment to be conducted with only a fraction of all of the possible combinations of parameter values.

The Taguchi method is a special variant of the design of experiments technique that distinguishes itself from classic DOE with a focus on optimizing design parameters to minimize variation in output parameters. Based on Taguchi's method and other DOE methodologies, Six Sigma's robust design performs experiments to investigate products where the output depends on many factors (variables, inputs) without having to tediously and uneconomically run experiments using all the variables' possible combinations of values. By systematically choosing certain combinations, it is possible to separate variables' individual effects.

As illustrated in Figure 3-5, orthogonal arrays, which specify the test cases to use to conduct the experiment, are used to aid in the design of an experiment. Frequently, two orthogonal arrays are used: a design factor matrix and a noise factor matrix. The latter is used to conduct an experiment in the presence of a difficult-to-control variation in order to develop a robust design.

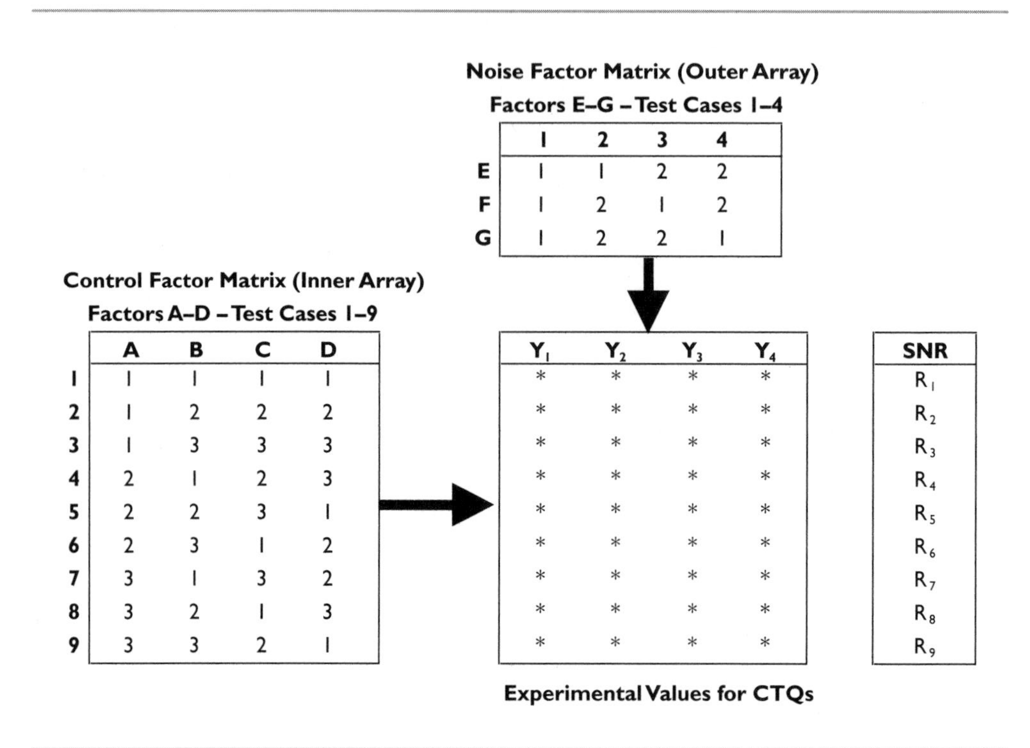

Figure 3-5 A design of experiments matrix.

This approach to designing and conducting an experiment to determine the effect of design factors (parameters) and noise factors on a performance characteristic is represented in the figure. These experimental results can be summarized into a metric called the *signal-to-noise ratio* that jointly considers how effectively the mean value (signal) of the parameter has been achieved and the amount of variability that has been experienced. As a result, a designer can identify the parameters that will have the greatest effect on the achievement of a product's performance characteristic.

The design parameters or factors of concern are identified in an inner array, or design factor matrix, that specifies the factor level or design parameter test cases. The outer array, or noise factor matrix, specifies the noise factor or the range of variation the product will be exposed to during the manufacturing process, in the environment, or by how the product is to be used (conditions it is exposed to). This experimental set-up allows the identification of the design parameter values, or factor levels, that will produce the best-performing, most reliable, or most satisfactory product over the expected range of noise factors or environmental conditions.

After experiments are conducted and the signal-to-noise ratio (SNR) is determined for each design factor test case, a mean SNR value is calculated for each design factor level or value. This data is statistically analyzed using analysis of variance (ANOVA) techniques, which are discussed in detail in Chapter 4.

Very simply, a design factor with a large difference in the signal-to-noise ratio from one factor setting to another indicates that the factor or design parameter is a significant contributor to the achievement of the performance characteristic. When there is little difference in the SNR from one factor setting to another, this indicates that the factor is insignificant with respect to the performance characteristic.

With the understanding resulting from the experiments and subsequent analysis, the designer can identify the following:

- Parameter values that maximize achievement of a performance characteristic and minimize the effect of noise, thereby achieving a more robust design
- Parameters that have no significant effect on performance; in those cases, tolerances can be relaxed and costs reduced
- Parameter values that reduce cost without affecting performance or variation

These steps take initial effort, but they can reduce cost and improve the performance of the product. In the past, the designer selected parameters and tolerances and made system design tradeoffs in an intuitive manner, sometimes supported by limited analysis and trial-and-error experimentation. However, an overall framework was lacking to make these decisions. DOE techniques offer a way to develop a more rigorous understanding of the relationship between product and process parameters and the achievement of a performance, reliability, and/or quality characteristics, thereby leading to improved design decisions. The technique represents a comprehensive approach to experimental design, analysis, and product and process design decision making.

3.5 EXPERIMENTS FOR DEPLOYING ROBUST QUALITY FUNCTIONS

To deploy a robust quality function for product development, a Six Sigma project team needs to investigate available testing options. The team must determine if the experiment will be done on hardware, with a test setup, or by simulation, and then acquire the necessary resources. It is often less costly to run experiments on test fixtures. Experimenting on a test fixture has the potential of producing generic results that can be used across production lines.

Simulating an experiment using a mathematical model is a very efficient means of testing. Parameter Design experiments require testing design combinations at the established noise conditions, so noise must be included in the math model. Unfortunately, this is uncommon.

Operational definitions are precise statements about the characteristics and procedures involved in the experimental process. It is important to ensure that those involved in both planning and running the experiment have a common understanding of all operational definitions, so establish precise ones for the following:

- Control factors
- Control factor levels and how to set them
- Noise factors
- Noise factor levels and/or conditions and how to set them
- CTQ characteristics and how to measure them

Before conducting the entire experiment, you should test the existing design, or a prototype (if available) several times at each noise level. This will serve three purposes: to test the noise strategy, to test the procedure, and to observe variation. The noise strategy should be strong enough to overwhelm DOE errors. For example, experimental design can be used to sequester the sources of undesired variability, making it possible to substantially reduce noise. As a result, proven statistical methodologies such as ANOVA can be used to take away variability due to identifiable nuisance effects from noise. If an unanticipated amount of variation within each noise setting is observed, attempt to ascertain the source or sources (e.g., poor measurement capability, environmental conditions). The team should incorporate any overwhelming noise that may be present into the noise strategy; modify the experimental procedure; or, if necessary, consider replication or randomization to minimize the impact of the error.

Because only one or two designs are examined during preliminary testing, repetition is recommended. The better the noise strategy and the better the repeatability, the fewer repetitions that will be needed in the orthogonal array experiment. As shown in Figure 3-6, a dot plot of the results can be used to verify that the noise strategy separates the responses into observable subgroups.

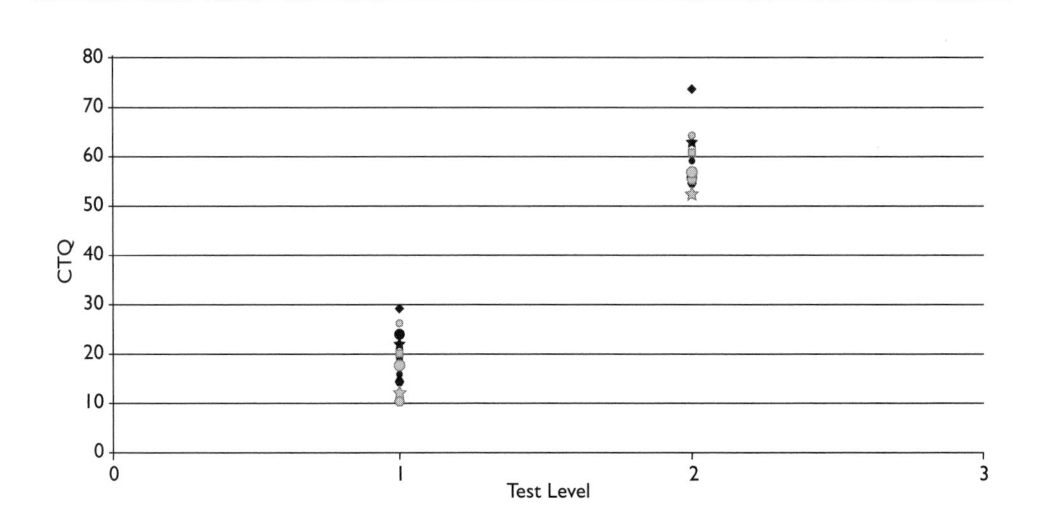

Figure 3-6 Use a dot plot for preliminary testing.

EXAMPLE 3.1

A ball-swirling machine experiment is to be conducted on one of the machines in the plant. Based on preliminary testing and on knowledge of the system, the team is confident that, on average, the noise strategy provides adequate separation relative to common causes of error. But, due to sporadic lack of reproducibility and in consideration of the minimal cost implications, they decide to collect three swirl times at each noise level for each design. Here are the procedures.

1. *Swirl time (CTQ):* Begin timing when the timer hears the ball enter the funnel and end when you hear the ball land in the receptacle. CTQs are the key measurable characteristics of a product whose performance standards or specification limits must be met in order to satisfy the customer. They align improvement or design efforts with customer requirements.

2. *Procedure for measuring CTQ:* Create a "gate" on the ramp (at the appropriate run-length level) using the end of a ruler. Place the ball behind the gate. When the timer is ready, count down—"Ready, set, start"—and release the ball by lifting the gate.

As shown in Table 3-1, the following are the control factors:

- *Run length*—the distance, in millimeters (mm), from the end of the ramp entering the funnel to the position from which the ball is released; mark the inside of the ramp at the lengths indicated.
- *Ramp-to-funnel angle*—the angle formed by the centerline of the funnel and the ramp; fix the centerline of the funnel, and draw lines on the table to denote the two angle levels.
- *Run end height*—the distance, in millimeters, between the end of the ramp and the table.

Table 3-1 Control and Noise Factors for the Ball-Swirling Machine

Control Factors (for inner array)	**Level 1**	**Level 2**
L: Run length	900mm	600mm
A: Ramp-to-funnel angle	30 degrees	45 degrees
H: Run end height	500mm	600mm
C: Clamping (unscrewed)	0.0 turns	0.5 turns
O: Operator training	Yes	No
Noise Factor (for outer array)		
Ball size	9mm	15mm

- *Clamping*—the degree of clamping tightness of the upper clamp, where 0 indicates fully tightened and 0.5 indicates half a revolution of the clamping screw.
- *Operator training*—identify one operator who has had no training and a minimal amount of experience and one with training and lots of experience. Each operator should acknowledge the dropping technique they typically employ and should maintain that technique throughout the experiment. Do not substitute alternate operators during the course of the experiment.

Here, ball bearing size is selected as a noise factor. The size is measured by the diameter of the ball bearings provided by supplier X. Use the same two ball bearings for the entire experiment.

The team should now be in the position to execute the experiment. Frequently, it is difficult, or expensive, to change either the noise conditions or the designs from run to run. Generally, it is a good idea to collect data in the order that requires the fewest number of changes in the characteristic that is the most difficult to reset.

For example, if it is difficult or expensive to change the noise conditions, first collect all of the data (for all runs) at the N_1 condition. Then change to the N_2 conditions and collect the remaining data (see Figure 3-7).

To maintain balance, all designs specified in the inner array must be tested, which is critical to deploying quality functions. Without balance, the ensuing analysis approach will be invalid because all of the Quality Function Deployments will be biased.

Run No.	A	B	C	...
1				
2				
3				
•				
•				
•				

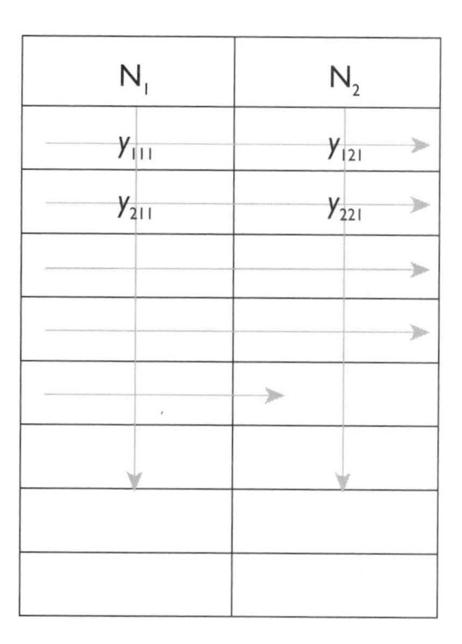

Figure 3-7 Order of data collection.

3.6 SUMMARY

Quality Function Deployment is a set of powerful product development tools that identify the Critical-to-Quality characteristics during the new product development process. The main features of QFD are a focus on meeting customer requirements based on the VOC models described in Chapter 2. Through cross-functional teamwork, a Six Sigma team establishes the House of Quality—a comprehensive matrix for documenting information, specifications, and decisions. QFD helps to reduce the time to market, minimize design changes, improve quality, and increase customer satisfaction with new products.

Design of experiment methods are widely used in industrial settings for deploying quality functions. Since industrial experiments are expensive, the primary objective is usually to extract the maximum amount of unbiased information regarding the control factors affecting a product design from as few experiments as possible. Deploying quality functions, DOE optimizes the selection of control factors early in the product development process. The correlation and contraction of the control factors are the basis for generating effective design concepts using the Theory of Innovative Problem Solving (TRIZ) presented in Chapter 4.

BIBLIOGRAPHY

Clausing, D. P. *Total Quality Development: A Step-by-Step Guide to World-Class Concurrent Engineering.* New York: ASME Press, 1994.

Cohen, L. *Quality Function Deployment, First Edition.* Englewood Cliffs, NJ: Prentice Hall PTR, 1995.

Madu, C. N. *House of Quality (QFD) in a Minute.* Fairfield, CT: Chi Publishers, 2000.

Pahl, G., and W. Beitz. *Engineering Design: A Systematic Approach.* New York: Springer, 1988.

Pugh, Stuart. *Total Design: Integrated Methods for Successful Product Engineering.* Wokingham, UK: Addison Wesley, 1991.

Roush, M. L., M. Modarres, and R. N. Hunt. "Application of Goal Tree to Evaluation of the Impact of Information upon Plant Availability." Proceedings of the International ANS/ENS Topical Meeting on Probabilistic Safety Methods and Applications, San Francisco, 1985.

Taguchi, G. *The System of Experimental Design: Engineering Methods to Optimize Quality and Minimize Costs.* Dearborn, MI: Quality Resources, 1987.

Terninko, J. *Step-by-Step QFD: Customer-Driven Product Design, Second Edition.* Delray Beach, FL: Saint Lucie Press, 1997.

Terninko, J. "The QFD, TRIZ and Taguchi Connection: Customer-Driven Robust Innovation." Proceedings of the Ninth Symposium on Quality Function Deployment, Novi, MI, June 10, 1997.

4 Theory of Inventive Problem Solving

Creating Robust Design Concepts

Technical contradictions between Critical-to-Quality (CTQ) characteristics account for one of the main difficulties when developing robust designs. For example, a product development team at Micro Electro-Mechanical Systems (MEMS) is designing a crankshaft sealing system for a micro engine. Enhancing the sealing performance requires a tight contact between the rotating crankshaft and the stationary seal. However, a tight contact will intensify friction, accelerate mechanical wear-out, and thus jeopardize system reliability.

The Theory of Inventive Problem Solving (TRIZ, its Russian acronym) helps generate robust design concepts for eliminating technical contradictions. Integrating value-added features enables engineers to determine the final robust design concept. As this chapter illustrates with a practical example, the quality of the design concept is the design's *DNA*, which drives robustness. A process for synthesizing ideas into a superior concept for development is also included here. Based on TRIZ, it is possible to perform various Six Sigma design activities, including the ones that follow.

- Select the design concept systematically with a *Pugh matrix*, which is a scoring tool used for concept selection—the options of which are assigned scores relative to criteria. Selection is made based on the consolidated scores. Before starting a detailed design, you need to have many options so that the best of them can be chosen.
- Use *t*- and *F*-tests to compare two design concepts. A *t-test* is a statistical tool used to determine whether a significant difference exists between the means of two distributions or the mean of one distribution and a target value. An *F-test* is used to determine whether two samples that are drawn from different populations have the same standard deviation with a specified confidence level (samples may be of different sizes).
- Use an analysis of variance (ANOVA) to compare three or more design concepts. *ANOVA* is a calculation procedure to allocate the amount of variation in a process and to determine whether it is significant or is caused by random noise.

Russian scientist and writer Genrich Saulovich Altshuller developed TRIZ. Faced with the challenges of inventing, Altshuller searched scientific literature for a systematic method to do so; he believed that one must exist. To his disappointment, he could not find any clues as to the existence of any method. He came to the conclusion that he needed to develop a method himself.

Altshuller started off by examining a large database of his own and other people's inventions and soon arrived at his most important finding: *Inventing* is the removal of a contradiction with the help of certain principles. To develop a method for inventing, he argued, one must scan a large number of inventions, identify the contradictions underlying them, and formulate the principle used by the inventors for their removal.

4.1 CONTRADICTION: THE GATEWAY TO NEW DESIGNS

A *contradiction* is a situation in which an improvement made in one feature of a system directly leads to the deterioration of another. For example, a contradiction appears in the design of a lightbulb: If energetic efficiency is improved, the bulb's lifespan is shortened. The common cure for a contradiction found in a system is *compromise*—finding the best tradeoff between contradictory requirements.

According to Altshuller (1994), the removal of a contradiction means creating a better situation without resorting to a tradeoff. In many cases, a contradiction arises as a result of conflicting requirements from a certain physical property. For example, in the case of the lightbulb, its temperature must be very high to ensure efficiency and very low to ensure longer life. The way Altshuller states the problem—instead of stating the common question about how to find the optimal temperature—is as follows: How could the temperature be both high and low?

To elaborate on what Altshuller meant by *principle*, let's discuss another example. When designing the engine crankshaft, vehicle design engineers face a very difficult problem: During the vehicle's operation, oil can leak from the crankshaft during high-speed rotation. To prevent leakage, the seal must be in very close contact with the crankshaft. But that increases the possibility of mechanical wear-out, which is not acceptable for reliability. The contradiction here is clear: A crankshaft seal needs to be both close and loose. Eventually this problem was solved by using Altshuller's "taking out" principle (2000), which means trying to remove something from the system. The solution was to seal the crankshaft with a labyrinth. This noncontact labyrinth continually loses oil, collects oil, and then recirculates oil. The idea is that instead of fighting leakage, oil is simply collected and reused.

4.2 THIRTY-NINE ENGINEERING PARAMETERS TO STANDARDIZE CTQs

Broadly speaking, a contradiction is when two or more statements, ideas, or actions are seen as incompatible. One must, it seems, reject at least one of the ideas outright. In logic, *contradiction* is defined much more specifically, usually as the simultaneous assertion of a statement and its negation ("denial" can be used instead of "negation"). This, of course, assumes that *negation* has a nonproblematic definition.

In the words of Aristotle (1886), "One cannot say of something that it is and that it is not in the same respect and at the same time" (Cooper, 1962). A given when engineering products is that customers often have contradicting requirements. Because of this, designs for products need to address contradictory CTQ characteristics.

The benefit of analyzing a particular innovative problem to find the contradictions is that TRIZ patent-based research directly links the type of contradiction to the most probable principles for solution of that problem. In other words, the general TRIZ model shown in Figure 4-1 is particularly easy to apply for contradictions.

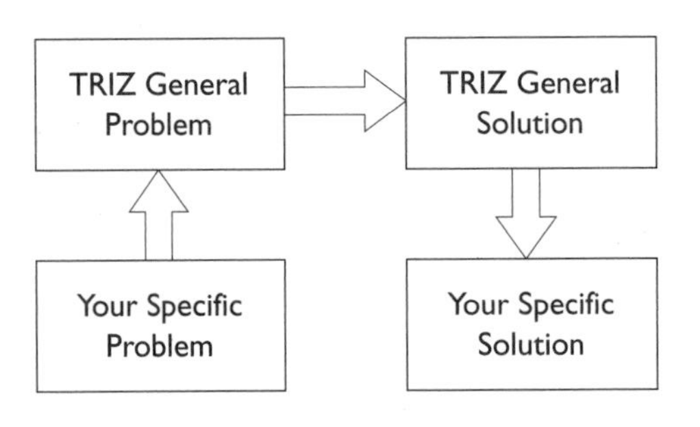

Figure 4-1 General TRIZ model for solving problems.

TRIZ defines two kinds of contradictions: technical and physical. *Technical* contradictions are the classic engineering tradeoffs. The desired state cannot be reached because something else in the system prevents it. In other words, when something gets better, something else gets worse. Here are a couple of classic examples:

- The product gets stronger (good) but the weight increases (bad)
- The bandwidth increases (good) but requires more power (bad)

Physical contradictions are situations in which one object has contradictory, or opposite, requirements. Everyday examples abound, including:

- When pouring hot filling into chocolate candy shells, the filling should be hot enough to pour fast but cool enough to prevent melting the chocolate.
- Software should be easy to use but should have many complex features and options.

A basic principle of TRIZ is that a technical problem is defined by contradictions; that is, if there are no contradictions, there are no problems. This radical-sounding statement forms the basis for the fastest and easiest-to-learn TRIZ problem-solving methods. The contradiction matrix is one of the effective and visual tools of TRIZ.

The idea of the matrix—and its initial development—was created and worked out by Altshuller in the process of researching about 40,000 patents with so-called "strong levels of solutions" in several classes of the International Patent Classification system. This database was extracted from nearly 400,000 worldwide patents during the sixties and seventies.

Altshuller figured out 39 types of 39 engineering parameters on the matrix's axes—"Undesired Effect" versus "Feature to Improve"—and

40 types of Inventive Principles. The 39 engineering parameters are shown in Table 4.1. Fundamentally, robust design is concerned with minimizing the effect of uncertainty or variation in engineering parameters on a design without eliminating the source of the uncertainty or variation.

EXAMPLE 4.1

This example uses a beverage can to illustrate the TRIZ problem-solving approach (Altshuller 2000, Liu Chen 2001). A beverage can is an engineered system to contain a drink. The can, a universally accepted container for many foods, came tip-toeing into the marketplace as a substitute for the bottling of beverages. Unlike the bottle, the can could be made in many shapes and designs, and beverage makers like the ability to use the whole can's surface to promote brand recognition. A much sturdier container than that used for food products is required to withstand the 80 to 90 psi pressure of pasteurization, in contrast to the 25 to 30 psi used during food processing.

The operating environment is that cans are stacked for storage purposes. As demand for soft drinks in cans grows, filling speed becomes an important factor. Sources of information to consider include weight of filled cans, internal pressure of can, rigidity of can construction. Its primary useful function is to contain beverages. Harmful effects include cost of materials to produce cans and waste of storage space. The ideal result is a can that will support the weight when stacked to human height without damaging cans or the beverages in them.

4.2.1 IDENTIFYING CONTRADICTIONS

The height to which cans will be stacked cannot be controlled. The price of raw materials compels manufacturers to lower costs, so engineers are designing cans to use the thinnest aluminum possible within the constraints of strength and appearance. The can walls must be made thinner to

Table 4-1 The 39 Engineering Parameters

1. Weight of moving object	21. Power
2. Weight of nonmoving object	22. Waste of energy
3. Length of moving object	23. Waste of substance
4. Length of nonmoving object	24. Loss of information
5. Area of moving object	25. Waste of time
6. Area of nonmoving object	26. Amount of substance
7. Volume of moving object	27. Reliability
8. Volume of nonmoving object	28. Accuracy of measurement
9. Speed	29. Accuracy of manufacturing
10. Force	30. Harmful factors acting on object
11. Tension, pressure	31. Harmful side effects
12. Shape	32. Manufacturability
13. Stability of object	33. Convenience of use
14. Strength	34. Repairability
15. Durability of moving object	35. Adaptability
16. Durability of nonmoving object	36. Complexity of device
17. Temperature	37. Complexity of control
18. Brightness	38. Level of automation
19. Energy spent by moving object	39. Productivity
20. Energy spent by nonmoving object	

reduce costs but if walls are made thinner, cans cannot support as large a stacked load. Thus, the can wall needs to be thinner to lower material cost and thicker to support stacking-load weight: This is a physical contradiction. If a solution can be found, an ideal engineering system will be achieved. In the 1970s, the aluminum in beverage cans was nearly as thick as aluminum gutters—0.015 inches. The progress in reducing weight leveled off in the early eighties, then resumed in 1984 due in part to computer modeling.

4.2.2 Using Standard Engineering Parameters

The standard engineering parameter that has to be changed to make the can wall thinner is #4, "length of a nonmoving object." In TRIZ, these standard engineering principles can be quite general. Here, "length" can refer to any linear dimension such as length, width, height, diameter, and so on. If the can wall is made thinner, it becomes structurally weaker. The standard engineering parameter that is in conflict is #11, "tension, pressure." The standard technical conflict is: The more one improves the standard engineering parameter—length of a nonmoving object—the more the standard engineering parameter—tension, pressure—becomes worse. Finite-element modeling of beverage cans allows engineers to develop prototype can designs with a high level of confidence so that the modeling will accurately predict how the can will perform under the stress of manufacturing, distribution, and use.

4.3 Forty Principles to Identify Design Solutions

One of the central ideas of TRIZ is a design analysis to find areas of conflict between improving and worsening features, the outcome being a predictable subset of Inventive Principles that resolve that conflict. For instance, you may have a situation where bright light is a necessary function but has a downside of fading the material that it shines on. Cross-referencing "illumination intensity" against "stability of composition" on the TRIZ contradiction matrix immediately returns ready-made resolutions: optimal property changes (i.e., shielding), nonuniform material (i.e., local shielding), and disposability (i.e., let the fading happen, but replace the component).

The 40 principles are the easiest TRIZ tool to use, and the one most likely to give you good solutions fairly easily and quickly. They are the ways the world has found to solve contradictions. TRIZ researchers discovered that the most powerful solutions (innovative solutions) are the ones in which contradictions are eliminated and that they only occurred in about 20 percent of the world's successful patents that were analyzed. When studying which patents uncovered and overcame a contradiction, researchers recorded all the ways used; to their astonishment, they found just 40 different strategies—even after examining more than half a million innovative patents.

In the West today, most engineers are taken aback to find there are such simple, easy-to-learn rules for solving problems. The 40 Inventive Principles can be used to help solve both technical and nontechnical problems quickly and simply. They show you how to tap into others' work and good results; and, with the contradiction matrix and the principles, it is clear that the TRIZ tool has been used successfully to eliminate contradictions and find powerful solutions.

The important and exciting reality is that there are only 40. The fastest way of getting started with TRIZ is to become familiar with each one and to use them in conjunction with whatever system you normally use to solve problems. The following pages describe several of the principles and give simple examples of where they are used in real life, at home and at work. Remember each example demonstrates how a contradiction has been solved and that it is always worth looking for that contradiction. All 40 Inventive Principles are listed with examples in this chapter's appendix.

Engineers need to develop ultrareliable and yet low-cost, spacious, compact, powerful, and energy-saving products. An engineering design is the result of contradiction resolution defined not by indifference of the two design parameters in their distinction, but by their being bound together in one unit. In engineering design, this is often called the "law of contradiction," or the "law of the unity of contradicting CTQs"; it is a basic law used when engineering robust designs with Six Sigma. The Table of Contradictions in Altshuller (1984) summarizes how to apply the 40 engineering problem-solving principles to resolve contradictions in production designs.

EXAMPLE 4.1 CONTINUED

The engineering parameters in conflict for the beverage can are #4, length of a nonmoving object; and #11, tension, pressure. The feature to improve (Y-axis) is the can wall thickness, or #4, and the undesirable secondary effect (X-axis) is loss of load-bearing capacity, or #11. By looking at Altshuller's Table of Contradictions in his book, *Creativity as an Exact Science* (1984), you can see that numbers 1, 14, and 35 are in an intersecting cell.

4.3.1 PRINCIPLE 1. SEGMENTATION

This is a process used to divide a large group into smaller, logical categories for analysis. Some commonly segmented entities are customers, datasets, or markets.

A. Divide an object into independent parts.
- Replace mainframe computer by personal computers.
- Replace a large truck by a truck and trailer.
- Use a work breakdown structure for a large project.

B. Make an object easy to disassemble.

- Rapid-release fasteners for bicycle saddle, wheel, and so on.
- Quick-disconnect joints in plumbing and hydraulic systems.
- Single fastener V-band clamps on flange joints.
- Looseleaf paper in a ring binder.

C. Increase the degree of fragmentation or segmentation.

- Multiple control surfaces on aerodynamic structures.
- Internal combustion engine valves of 16 and 24 versus 8.
- Multizone combustion system.
- Build up a component from layers (e.g., stereo-lithography, welds, etc.).

In robust design, segmentation involves dividing product data into parts. For example, you may collect the product's cause of defect and display the segmentation data in these possible ways:

- Type A defects are 50 percent
- Type B defects are 30 percent
- Type C defects are 10 percent

Design Solution Based on Principle 1

Using Inventive Principle 1c—"Increase the degree of an object's segmentation"—the wall of the can could be changed from one smooth continuous wall to a corrugated or wavy surface made up of many little walls. This would increase edge strength yet allow a thinner material to be used (see Figure 4-2).

Good sheet formability is required for the body-making process, which involves blanking, cupping, and finally drawing and ironing the side walls. Anisotropy in the mechanical behavior of the sheet must be minimized to

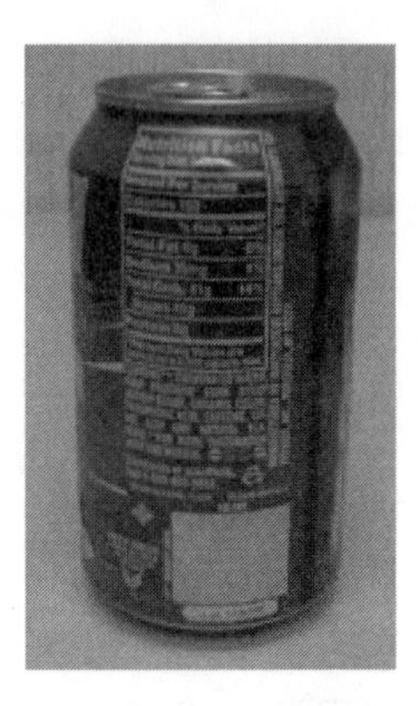

Figure 4-2 Can a corrugated wall be made using Principle 1?

limit the formation of so-called "ears" on the deep drawn cup. The can body needs to be able to withstand a minimum dome-reversal pressure and have vertical load-bearing capacity, making strength an important consideration also. Higher strength enables thinner sheets to be used and hence allows more efficient use of materials during product fabrication.

4.3.2 PRINCIPLE 14. SPHEROIDALITY—CURVATURE

Spheroidality means replacing flat surfaces with curved ones or replacing cubical shapes with spherical shapes. For example, use arches and domes for strength in bridge construction and building architecture. More application examples include:

A. Instead of using rectilinear parts, surfaces, or forms, use curvilinear ones; move from flat surfaces to spherical ones; from parts shaped as a cube (parallelepiped) to ball-shaped structures.

- Use arches and domes for strength in architecture.
- Introduce fillet radii between surfaces at different angles.
- Introduce stress-relieving holes at the ends of slots.
- Change curvature on lens to alter light-deflection properties.

B. Use rollers, balls, spirals, domes.

- Spiral gear (e.g., Nautilus) produces continuous resistance for weight lifting.
- Ballpoint and roller-point pens facilitate smooth ink distribution.
- Use spherical casters instead of cylindrical wheels to move furniture.
- Archimedes screws are useful.

C. Go from linear to rotary motion; use centrifugal forces.

- Rotary actuators in hydraulic system.
- Switch from reciprocating to rotary pump.
- Push/pull versus rotary switches (e.g., lighting dimmer switch).
- Linear motors.
- A computer mouse makes use of a ball structure to convert linear two-axis motion into vector motion.
- Screw-thread versus nail.

Design Solution Based on Principle 14

The aluminum beverage can is made with two pieces—the can body and the can end (or lid). The manufacturing process starts with coils of aluminum. Can-making plants use massive quantities of aluminum coil every day to make can bodies or ends. Each coil typically weighs about 25,000 pounds and, when rolled out flat, can be anywhere from 20,000-feet to 30,000-feet long and five- to six-feet wide. Using Inventive Principle 14a, the perpendicular angle at which most can lids are welded to the can wall can be changed to a curve (see Figure 4-3). The can end is made by blanking, drawing, curl forming, riveting, and producing a score line for the

Figure 4-3 Curved can-lid welding joint using Principle 14a.

easy-open tab end. As well as high strength, stock for can ends needs good formability and good surface finish (no streaking). Based on Principle 14, the top of the aluminum beverage can is assembled with curved lines instead of straight ones.

4.3.3 PRINCIPLE 35. PARAMETER CHANGES

Parameter change means transforming an object's aggregate state, density distribution, degree of flexibility, or temperature. This brings to mind water; for example, water in the form of a liquid in some cases and in the form of solid ice in other cases, denoting changes in properties due to the physical-phase transition. Another example is the problem that arises when making chocolate pieces that contain artificial alcohol flavoring. The flavoring needs to be inside in the form of liquid, but it can't be dispensed inside the chocolate. One of the standard solutions is freezing the artificial alcohol flavoring in molds and simply putting it into chocolate in the form of solid ice. The following are more application examples.

A. Change an object's physical state (e.g., to a gas, liquid, or solid).

- Transport oxygen, nitrogen, or petroleum gas as a liquid, instead of a gas, to reduce volume.
- Freeze the liquid centers of filled candies, then dip in melted chocolate instead of handling the messy, gooey, hot liquid.

B. Change the concentration or consistency.

- Liquid soap.
- Abradable linings used for gas-turbine engine seals.

C. Change the degree of flexibility.

- Vulcanize rubber to change its flexibility and durability.
- Use compliant brush seals rather than labyrinth or other fixed geometry seals.

D. Change the temperature.

- Raise the temperature above the Curie point to change a ferromagnetic substance to a paramagnetic substance.

Design Solution Based on Principle 35

Change the can wall composition to a stronger metal alloy to increase the load-bearing capacity. Typically, the aluminum for cans contains 1 percent Mg, 1 percent Mn, 0.4 percent Fe, 0.2 percent Si, and 0.15 percent Cu. This kind of alloy is both very ductile before drawing and strong when drawn. The strength comes from the fact that the Mg and Mn atoms are slightly larger than the Al atoms. The iron and manganese also produce tiny particles with each other and the aluminum during the deformation, adding strength.

After manufacture, can bodies and ends are transported to a filling plant where the beverage is put in and the two components are attached using a folded seam and a small amount of sealing compound. Thicker and stronger aluminum is needed for the lid, so the alloy used has less Mn and more Mg. The rivet in the center of the lid is made from the lid, not added. It is drawn; that is, it is punched from the lid and flattened to hold the tab, which serves as a tool to press open the portion of the scored lid that provides an opening so that the beverage can be drunk.

The shape of the bottom is another parameter to be optimized (Figure 9-4). While the top has to be flat, the bottom is domed to resist the pressure of the beverage using a thin metal wall. Both the top and bottom are designed to withstand about 100 psi. Optimization, involving Parameter Design (see Chapter 7) and Tolerance Design (see Chapter 8), means adjusting product inputs to produce the best possible average response with minimum variability.

Through the use of TRIZ, new design concepts to remove contradictions between CTQs can be generated. To optimize each design concept, analyze data and then select the best overall design. This starts with calculating an average for the control factor effects on CTQs.

Figure 4-4 Optimize the shape of the can bottom using Principle 35.

4.4 CALCULATING THE AVERAGE OF CONTROL FACTOR EFFECTS

Engineers make decisions based on data. An important step in designing robust products with Six Sigma is to analyze data and choose optimal designs. It is important to do the following:

- Calculate control factor effects on the mean
- Calculate noise factor effects
- Construct a noise-by-control interaction plot
- Construct a response plot
- Construct a noise-by-control plot
- Use two-step optimization
- Calculate a signal-to-noise ratio for static responses
- Construct a signal-to-noise response plot

Figure 4-5 shows the data collected in the ball-swirling machine experiment. It is a good idea to look at the data before beginning the analysis to see if there are any peculiarities. Look for outliers, trends, and noise separation. Proceed with analysis, but keep in mind any observations made at this juncture.

If any values are clearly not feasible, these should be replaced with new data. However, do not omit extreme values that are within the realm of feasibility—some designs will inherently demonstrate a higher degree of variability than others. For example, in Run 1 at N_2, a recorded value of 130 seconds is clearly not feasible, whereas 15 seconds is. Outliers should never be replaced unless there is a strong rationale for so doing. This is a good reason to provide the test operator with room on the data-collection sheet for notes.

Run No.	A	H	O	C	L	Noise					
						N_1			N_2		
1	1	1	1	1	1	8.3	8.4	8.1	13.0	13.0	13.5
2	1	1	1	2	2	5.4	5.1	5.4	6.7	6.5	6.2
3	1	2	2	1	2	8.5	6.5	6.5	11.0	11.2	11.2
4	1	2	2	2	1	8.5	8.8	8.2	11.4	11.5	11.4
5	2	1	2	1	2	7.5	7.2	7.8	11.5	11.6	12.0
6	2	1	2	2	1	9.0	9.4	9.3	11.7	11.9	11.7
7	2	2	1	1	1	8.9	10.0	10.0	15.5	15.4	15.1
8	2	2	1	2	2	7.4	7.9	7.7	10.3	12.5	10.7

Figure 4-5 Collect data for ball-swirling machine.

If there appears to be a trend in the data, such as monotonicity in the response or the variability, consider potential sources. For example, if testing was conducted on a test fixture, were components of the fixture deteriorating or loosening over time? If the trend is strong and data collection was not randomized, the data could be corrupted.

To improve robustness to noise, the noise strategy must produce separation in the data. If all runs demonstrate approximately the same degree of separation, the team is unlikely to find any factors that will improve robustness.

As shown in Figure 4-6, the analysis begins with the calculation of control factors' effects on the mean, which is the average data point value within a dataset. To calculate the mean, add all of the individual data points then

Run No.	A	H	O	C	L	Noise						Average		
						N_1			N_2			N_1	N_2	*Y*
1	1	1	1	1	1	**8.3**	**8.4**	**8.1**	**13.0**	**13.0**	**13.5**	**8.3**	**13.2**	***10.7***
2	1	1	1	2	2	*5.4*	*5.1*	*5.4*	6.7	6.5	6.2	5.3	6.5	*5.9*
3	1	2	2	1	2	*8.5*	*6.5*	*6.5*	11.0	11.2	11.2	7.2	11.1	*9.2*
4	1	2	2	2	1	*8.5*	*8.8*	*8.2*	11.4	11.5	11.4	8.5	11.4	*10.0*
5	2	1	2	1	2	*7.5*	*7.2*	*7.8*	11.5	11.6	12.0	7.5	11.7	*9.6*
6	2	1	2	2	1	*9.0*	*9.4*	*9.3*	11.7	11.9	11.7	9.2	11.8	*10.5*
7	2	2	1	1	1	*8.9*	*10.0*	*10.0*	15.5	15.4	15.1	9.6	15.3	*12.5*
8	2	2	1	2	2	*7.4*	*7.9*	*7.7*	10.3	12.5	10.7	7.7	11.2	*9.4*

Figure 4-6 Calculating control factor effects.

divide that figure by the total number of data points. A factor with a large mean effect can be used to shift the mean of the system toward the target. To calculate the mean effects, first calculate the Average *Y*—the average of all six responses for each run.

The factors' effects on the mean can be calculated now (see Figure 4-7). Here the effect of a control factor is the difference between control factor level averages. For each factor, average the values in the Average *Y* column corresponding to Level 1. For example, for Factor A, average the Average *Y* values in runs 1 through 4. Next, average the values in the Average *Y* column corresponding to Level 2. For Factor A, average values 9.7, 10.5, 12.6, and 9.2. Finally, to calculate the effect, subtract the Level 2 average from the Level 1 average. So for Factor A, subtract 10.5 from 8.8, which results in a –1.6-second mean effect of Factor A. Repeat this process for all factors.

4.5 CALCULATING THE EFFECTS OF NOISE FACTORS

Now it is time to calculate N_1 and N_2 averages (Average N_1 and Average N_2) for each run. These are not used in the calculation of control factor effects on the mean but will be used later in the analysis.

Next, calculate the effect of noise. This will indicate the amount of variability, on the average, that was introduced into the experiment. The *noise* effect is the difference between the noise level averages.

Begin by calculating the average of all of the N_2 data, as highlighted in bold. If there is no missing data, averaging the values in the Average N_2 column will produce the same result (see Figure 4-8). Now average all of the N_1 values. Once again, if there is no missing data, averaging the values in the Average N_1 column will produce exactly the same result.

Run No.	A	H	O	C	L	Noise						Average		
						N_1			N_2			N_1	N_2	*Y*
1	**1**	1	1	1	1	*8.3*	*8.4*	*8.1*	13.0	13.0	13.5	8.3	*13.2*	***10.7***
2	**1**	1	1	2	2	*5.4*	*5.1*	*5.4*	6.7	6.5	6.2	5.3	6.5	**5.9**
3	**1**	2	2	1	2	*8.5*	*6.5*	*6.5*	11.0	11.2	11.2	7.2	11.1	**9.2**
4	**1**	2	2	2	1	*8.5*	*8.8*	*8.2*	11.4	11.5	11.4	8.5	11.4	***10.0***
5	2	1	2	1	2	*7.5*	*7.2*	*7.8*	11.5	11.6	12.0	7.5	11.7	*9.6*
6	2	1	2	2	1	*9.0*	*9.4*	*9.3*	11.7	11.9	11.7	9.2	11.8	*10.5*
7	2	2	1	1	1	*8.9*	*10.0*	*10.0*	15.5	15.4	15.1	9.6	15.3	*12.5*
8	2	2	1	2	2	*7.4*	*7.9*	*7.7*	10.3	12.5	10.7	7.7	11.2	*9.4*
Average 1	**8.9**	9.2	9.6	9.9	10.9									
Average 2	*10.5*	10.3	9.8	8.9	8.5									
Effect	−1.6	−1.1	−0.2	1.0	2.4									

Figure 4-7 Calculating factor effects on Average Y.

Run No.	A	H	O	C	L	Noise N_1			Noise N_2			Average N_1	Average N_2	Average Y
1	1	1	1	1	1	*8.3*	*8.4*	*8.1*	**13.0**	**13.0**	**13.5**	8.3	*13.2*	10.7
2	1	1	1	2	2	*5.4*	*5.1*	*5.4*	**6.7**	**6.5**	**6.2**	5.3	6.5	5.9
3	1	2	2	1	2	*8.5*	*6.5*	*6.5*	**11.0**	**11.2**	**11.2**	7.2	11.1	9.2
4	1	2	2	2	1	*8.5*	*8.8*	*8.2*	**11.4**	**11.5**	**11.4**	8.5	11.4	10.0
5	2	1	2	1	2	*7.5*	*7.2*	*7.8*	**11.5**	**11.6**	**12.0**	7.5	11.7	*9.6*
6	2	1	2	2	1	*9.0*	*9.4*	*9.3*	**11.7**	**11.9**	**11.7**	9.2	11.8	*10.5*
7	2	2	1	1	1	*8.9*	*10.0*	*10.0*	**15.5**	**15.4**	**15.1**	9.6	15.3	*12.5*
8	2	2	1	2	2	*7.4*	*7.9*	*7.7*	**10.3**	**12.5**	**10.7**	7.7	11.2	*9.4*
Average 1	8.9	9.2	9.6	9.9	10.9	**All N_2 Average**		**11.5**						
Average 2	10.5	10.3	9.8	8.9	8.5	**All N_1 Average**		7.9						
Effect	−1.6	−1.1	−0.2	1.0	2.4	**Effect**		3.6						

Figure 4-8 Calculating noise effects.

Subtract the All N_1 Average from the All N_2 Average to obtain the noise effect. A noise effect of 3.6 seconds implies that, on the average, there is a 3.6-second difference in swirl time between the 9mm ball bearing and the 14mm ball bearing.

As discussed in Chapter 3, the noise factors were allocated to the outer array so that the noise-by-control interaction effects, which are analyzed next, would be free from confounding.

4.6 Noise-by-Control Interaction Effects

An interaction occurs when the response achieved by one factor depends on the level of the other factor. When lines are not parallel on a plot, there's an interaction. To analyze noise-by-control interactions, all combinations of the interacting factor levels must be tested. Allocating the noise to the outer array has guaranteed this.

The procedure for calculating the noise-by-control effects is illustrated in Figure 4-9. For each control factor, to calculate an interaction effect, first calculate the swirl-time average for each of the following four control factor (CF) and noise (N) level combinations: CF_1 N_1, CF_1 N_2, CF_2 N_1, and CF_2 N_2. For example, for the noise-by-angle interaction effect, average the swirl-time averages corresponding to A at Level 1 and N at Level 1.

Then average the swirl-time averages corresponding to A at Level 2 and N at Level 1. Next, average the swirl-time averages corresponding to A at Level 1 and N at Level 2. Finally, average the swirl-time averages corresponding to A at Level 2 and N at Level 2. Using these four-level combination averages, the interaction effect can be calculated.

Chapter 3 showed the "longhand" formula for calculating interaction effects. As shown in Figure 4-9, this formula, with Levels 1 and 2 substituted

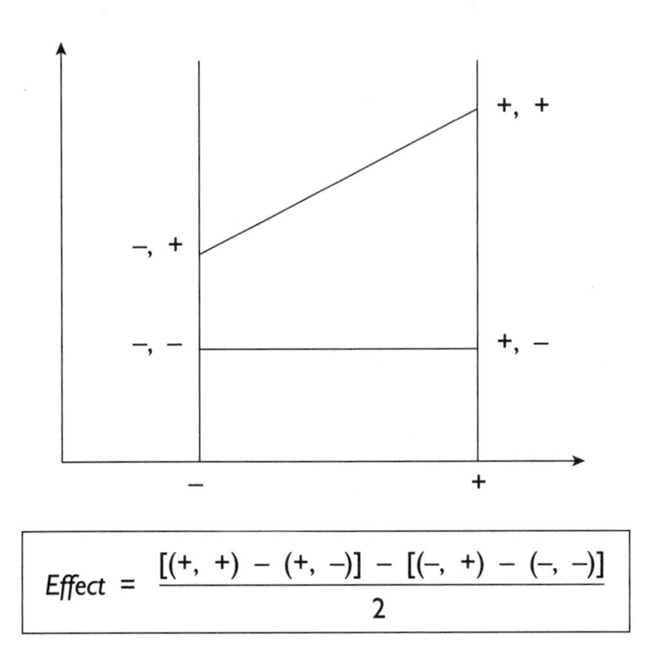

Figure 4-9 Calculating noise-by-control interaction effects.

for minus (–) and plus (+), will be used to calculate the noise-by-control interaction effects as follows:

$$
\begin{aligned}
\textit{Effect} &= \frac{[(+,\ +) - (+,\ -)] - [(-,\ +) - (-,\ -)]}{2} \\
&= \frac{[(1,\ 1) - (1,\ 2)] - [(2,\ 1) - (2,\ 2)]}{2} \\
&= \frac{[(7.1) - (10.6)] - [(8.6) - (12.4)]}{2} \\
&= 0.15
\end{aligned}
$$

The remaining noise-by-control interaction effects are calculated similarly; the results are shown in Figure 4-10.

Process input that consistently causes variation in the output measurement that is random and expected, and therefore not controlled, is called noise. Black noise is sources of variations that are nonrandom: that is, have a special cause. White noise is variations that are random or natural—a change in the source will not produce a predictable change in the response. Noise factors often conflict with the control factors for Critical-to-Quality characteristics.

In sum, control factors are design parameters whose nominal values can be cost-effectively adjusted by engineers. Noise factors are parameters

						Average	
Run No.	**A**	**H**	**O**	**C**	**L**	**N_1**	**N_2**
1	**1**	1	1	1	1	**8.3**	13.2
2	**1**	1	1	2	2	**5.3**	6.5
3	**1**	2	2	1	2	**7.2**	11.1
4	**1**	2	2	2	1	**8.5**	11.4
5	2	1	2	1	2	7.5	11.7
6	2	1	2	2	1	9.2	11.8
7	2	2	1	1	1	9.6	15.3
8	2	2	1	2	2	7.7	11.2

	A N_1	A N_2	H N_1	H N_2	O N_1	O N_2	C N_1	C N_2	L N_1	L N_2
CF1	7.3	10.6	7.6	10.8	7.7	11.5	8.1	12.8	8.9	12.9
CF2	8.5	12.5	8.2	12.3	8.1	11.5	7.7	10.2	6.9	10.1
Effect	0.15		0.41		–0.20		–1.08		–0.40	

Figure 4-10 Calculating noise-by-control interaction effects.

that influence system variability, but are difficult, impractical, or impossible to control. The appendix that follows summarizes the 40 Inventive Principles for removing contradictions and minimizing the noise-by-control interaction effects.

4.7 SUMMARY

The Theory of Inventive Problem Solving is an algorithmic approach for solving engineering problems. It is a scientifically based (as opposed to psychologically based) problem-solving process and toolkit for generating creative design concepts. Its foundation began with the study of worldwide patent literature and the classification of the most inventive patents. Analyzing these patents yielded the 40 generic Inventive Principles for solving and eliminating fundamental contradictions. These principles can be applied to all other engineering disciplines, greatly reducing the time to produce competitive ideas, concepts, and inventions during product development.

As revealed by the TRIZ approach, engineering robust designs involves fundamental contradictions. Most robust design solutions are tradeoffs. A desired improvement in a Critical-to-Quality design tradeoff often causes deterioration of another element of a CTQ characteristic. As a result, a product or system may not perform at least one of its functions perfectly; however, a creative solution to a difficult design problem satisfies conflicting CTQs—technical contradictions. Risks arising from contradictions can be managed by doing a Failure Mode and Effect Analysis, which is discussed in Chapter 5.

APPENDIX

FORTY PRINCIPLES TO DEAL WITH CONTRADICTIONS

Genrich Altshuller developed these 40 Inventive Principles more than 20 years ago. He and his team of associates reviewed thousands of world-wide patents selected specifically from leading industries for the inventive nature of their solutions to technical contradictions. In particular, he was interested in investigating contradictions that were resolved without compromise.

Altshuller found that utilizing principles previously used to solve similar problems in other inventive solutions could solve technical problems. For example, a "wearing problem" in the manufacturing of an abrasive product and a "wearing problem" with the cutting edge of a backhoe bucket were both solved using the Segmentation principle. Here the wearing problem is a noise exhibited in the manufacturing process. Altshuller's 40 Inventive Principles help to minimize the effects of noise-by-control interactions.

Principle 1. Segmentation

A. Divide an object into independent parts.

B. Make an object easy to disassemble.

C. Increase the degree of fragmentation or segmentation.

Principle 2. Taking Out

A. Separate an interfering part or property from an object, or single out the only necessary part (or property) of an object.

Principle 3. Local Quality

A. Change an object's structure from uniform to nonuniform; change an external environment (or external influence) from uniform to nonuniform.

B. Make each part of an object function in conditions that are the most suitable for its operation.

C. Make each part of an object fulfill a different and useful function.

Principle 4. Asymmetry

A. Change the shape of an object from symmetrical to asymmetrical.

B. If an object is asymmetrical, increase its degree of asymmetry.

Principle 5. Merging

A. Bring closer together (or merge) identical or similar objects; assemble identical or similar parts to perform parallel operations.

B. Make operations contiguous or parallel—bring them together in time.

Principle 6. Universality

A. Make a part or object perform multiple functions to eliminate the need for other parts.

Principle 7. "Nested Doll"

A. Place one object inside another; place each object, in turn, inside the other.

B. Make one part pass through a cavity in the other.

Principle 8. Antiweight

A. To compensate for the weight of an object, merge it with other objects that provide lift.

B. To compensate for the weight of an object, make it interact with the environment (e.g., use aerodynamic, hydrodynamic, buoyancy, and other forces).

Principle 9. Preliminary Antiaction

A. If it will be necessary to do an action with both harmful and useful effects, this action should be replaced with anti-actions to control harmful effects.

B. Create beforehand stresses in an object that will oppose known undesirable working stresses later on.

Principle 10. Preliminary Action

A. Perform, before it is needed, the required change of an object (either fully or partially).

B. Prearrange objects such that they can come into action from the most convenient place and without losing time for their delivery.

Principle 11. Beforehand Cushioning

A. Prepare emergency means beforehand to compensate for the relatively low reliability of an object.

Principle 12. Equipotentiality

A. In a potential field, limit position changes (e.g., change operating conditions to eliminate the need to raise or lower objects in a gravity field).

Principle 13. "The Other Way Around"

A. Invert the action(s) used to solve the problem (e.g., instead of cooling an object, heat it).

B. Make movable parts (or the external environment) fixed and fixed parts movable.

C. Turn the object (or process) upside down.

Principle 14. Spheroidality—Curvature

A. Instead of using rectilinear parts, surfaces, or forms, use curvilinear ones; move from flat surfaces to spherical ones; from parts shaped as a cube (parallelepiped) to ball-shaped structures.

B. Use rollers, balls, spirals, domes.

C. Go from linear to rotary motion; use centrifugal forces.

Principle 15. Dynamics

A. Allow (or design) the characteristics of an object, external environment, or process to change to be optimal or to find an optimal operating condition.

B. Divide an object into parts capable of movement relative to each other.

C. If an object (or process) is rigid or inflexible, make it movable or adaptive.

Principle 16. Partial or Excessive Actions

A. If 100 percent of an object is difficult to achieve using a given solution method then, by using "slightly less" or "slightly more" of the same method, the problem may be considerably easier to solve.

Principle 17. Another Dimension

A. To move an object in a two- or three-dimensional space.

B. Use a multistory arrangement of objects instead of a single-story arrangement.

C. Tilt or reorient the object; lay it on its side.

D. Use "another side" of a given area.

Principle 18. Mechanical Vibration

A. Cause an object to oscillate or vibrate.

B. Increase its frequency (even up to the ultrasonic).

C. Use an object's resonant frequency.

D. Use piezoelectric vibrators instead of mechanical ones.

E. Use combined ultrasonic and electromagnetic field oscillations.

Principle 19. Periodic Action

A. Instead of continuous action, use periodic or pulsating actions.

B. If an action is already periodic, change the periodic magnitude or frequency.

C. Use pauses between impulses to perform a different action.

Principle 20. Continuity of Useful Action

A. Carry on work continuously; make all parts of an object work at full load, all the time.

B. Eliminate all idle or intermittent actions or work.

Principle 21. Skipping

A. Conduct a process, or certain stages (e.g., destructible, harmful, or hazardous operations), at high speed.

Principle 22. "Blessing in Disguise" or "Turn Lemons into Lemonade"

A. Use harmful factors (particularly, harmful effects of the environment or surroundings) to achieve a positive effect.

B. Eliminate the primary harmful action by adding it to another harmful action to resolve the problem.

Principle 23. Feedback

A. Introduce feedback (referring back, cross-checking) to improve a process or action.

B. If feedback is already used, change its magnitude or influence.

Principle 24. "Intermediary"

A. Use an intermediary carrier article or intermediary process.

B. Merge one object temporarily with another (which can be easily removed).

Principle 25. Self-service

A. Make an object serve itself by performing auxiliary helpful functions.

B. Use waste resources, energy, or substances.

Principle 26. Copying

A. Instead of an unavailable, expensive, fragile object, use simpler and inexpensive copies.

B. Replace an object, or process, with optical copies.

C. If visible optical copies are already used, move to infrared or ultraviolet copies.

Principle 27. Cheap Short-Living Objects

A. Replace an inexpensive object with a multiple of inexpensive objects, comprising certain qualities (e.g., service life).

Principle 28. Mechanics Substitution

A. Replace a mechanical means with a sensory (optical, acoustic, taste, or smell) means.

B. Use electric, magnetic, and electromagnetic fields to interact with the object.

C. Change from static to movable fields, from unstructured fields to those having structure.

D. Use fields in conjunction with field-activated (e.g., ferromagnetic) particles.

Principle 29. Pneumatics and Hydraulics

A. Use gas and liquid parts of an object instead of solid parts (e.g., inflatable, filled with liquids, air cushion, hydrostatic, hydroreactive).

Principle 30. Flexible Shells and Thin Films

A. Use flexible shells and thin films instead of three-dimensional structures.

B. Isolate the object from its external environment using flexible shells and thin films.

Principle 31. Porous Materials

A. Make an object porous or add porous elements (e.g., inserts, coatings, etc.).

B. If an object is already porous, use the pores to introduce a useful substance or function.

Principle 32. Color Changes

A. Change the color of an object or its external environment.

B. Change the transparency of an object or its external environment.

Principle 33. Homogeneity

A. Make objects interact with a given object of the same material (or material with identical properties).

Principle 34. Discarding and Recovering

A. Make portions of an object that have fulfilled their functions go away (e.g., discard by dissolving, evaporating, etc.) or modify these directly during operation.

B. Conversely, restore consumable parts of an object directly in operation.

Principle 35. Parameter Changes

A. Change an object's physical state (e.g., to a gas, liquid, or solid).

B. Change the concentration or consistency.

C. Change the degree of flexibility.

D. Change the temperature.

Principle 36. Phase Transitions

A. Use phenomena occurring during phase transitions (e.g., volume changes, loss or absorption of heat, etc.).

Principle 37. Thermal Expansion

A. Use thermal expansion (or contraction) of materials.

B. If thermal expansion is being used, use multiple materials with different coefficients of thermal expansion.

Principle 38. Strong Oxidants

A. Replace common air with oxygen-enriched air.

B. Replace enriched air with pure oxygen.

C. Expose air or oxygen to ionizing radiation.

D. Use ionized oxygen.

E. Replace ozonized (or ionized) oxygen with ozone.

Principle 39. Inert Atmosphere

A. Replace a normal environment with an inert one.

B. Add neutral parts, or inert additives, to an object.

Principle 40. Composite Materials

A. Change from uniform to composite (multiple) materials.

BIBLIOGRAPHY

Altshuller, G. S. *The Innovation Algorithm.* Worcester, MA: Technical Innovation Center, 2000.

Altshuller, G. S. *Creativity as an Exact Science.* New York: Gordon & Breach Science Publishing House, 1984.

Altshuller, Henry. *The Art of Inventing (and Suddenly the Inventor Appeared).* Trans. Lev Shulyak. Worcester, MA: Technical Innovation Center, 1994.

Clapp, T. "Integrating TRIZ-Based Methods into the Engineering Curriculum." Proceedings of the IMC Users Group Conference (pp. 18–23), London, 1998.

Cooper, Lane, trans. and intro. *The Rhetoric of Aristotle.* Englewood Cliffs, NJ: Prentice Hall, 1962 (1886 edition published by Macmillan).

Fey, V., E. Rivin, and I. Vertkin. "Application of the Theory of Inventive Problem Solving to Design and Manufacturing Systems." *Annals of the CIRP,* 43(1): 107–10, 1994.

Kamm, L. J. *Real-World Engineering*. Piscataway, NY: IEEE Press, 1991.

Liu, C., and J. Chen. "A TRIZ Inventive Design Method Without Contradiction Information." *The TRIZ Journal*, September 2001—*http://www.triz-journal.com.*

Rivin, E. "Use of the Theory of Inventive Problem Solving (TRIZ) in Design Curriculum." Innovations in Engineering Education, ABET Annual Meeting Proceedings (pp. 161–64), 1996.

Terninko, J., A. Zusman, and B. Zlotin. *Systematic Innovation*. Boca Raton, FL: Saint Lucie Press, 1998.

Terninko, J., A. Zusman, and B. Zlotin. *Step-by-Step TRIZ: Creating Innovative Solution Concepts*. Nottingham, NH: Responsible Management, 1996.

5 Failure Mode and Effect Analysis

Being Robust to Risk

The engineering of products presents risks while providing benefits. Risk is about potential failures when existing products are used or new products are developed. As a predictive measure, the concept of risk is very important to minimize liability and maximize robustness. Failure Mode and Effect Analysis (FMEA) helps to identify every possible failure mode of a product, process, or service and to determine its effect on subitems and on the required function of the product, process, or service. By doing an FMEA, engineers can establish design controls to minimize risks. Design controls are key elements for developing robust design strategies.

This chapter illustrates how to develop an FMEA and how to use one for engineering robust designs; it includes a sample Failure Mode and Effect Analysis. In addition, the chapter discusses how to model the cause–effect relationship for risk management, how to assess and manage the dominant failure modes proactively to ensure product delivery, and how to assess and manage risk to ensure the achievement of project benefits.

As a tool with its roots within the Six Sigma methodology, FMEA can help identify and eliminate concerns early in the development of a new product or in the delivery of a new service. It is a systematic method to examine a product for possible ways in which failure can occur, and then to redesign a new model of the product to eliminate the possibility of failure. Properly executed, an FMEA can significantly assist in improving overall customer satisfaction and safety levels.

5.1 THE HISTORICAL CONTEXT OF FMEA

The FMEA methodology was developed by the U.S. military and has been widely used in defense and commercial industries. Military Procedure MIL-P-1629—*Military Standard Procedures for Performing a Failure Mode, Effects and Criticality Analysis (FMECA)*—has been used as a reliability assessment technique to determine the effect of system and equipment failures. Failures were categorized according to their impact on mission success, personnel safety, equipment integrity, system performance, maintainability, and maintenance and/or service requirements. The MIL-P-1629 standard established requirements and procedures for performing an FMECA to systematically evaluate and document, by item, the potential impact of every functional failure. Each possible failure is ranked by the severity of its effect so that appropriate preventive actions can be taken to eliminate or control high-risk items, which is now becoming the central part of risk engineering and management (Wang and Roush, 2000).

As a quality tool designed to improve the process of discovering how systems fail, FMEA was used by the aerospace industry during the 1960s. Since then, this tool has been adopted by many diverse industries and its scope of application has expanded to include elements of failure prevention and risk management relating to areas such as design, service, and equipment.

One area where the use of FMEA analysis has been emphasized is in the maritime industry. For example, the design control of the marine electronic system to satisfy the demands of the sea can be reviewed and improved through an FMEA. As you no doubt know from the movie *Titanic,* safety is immensely important and is of the utmost concern when developing the designs and processes used to build ships, submarines, large vessels, and even recreational boats. The need for a better understanding of the safe performance of new marine designs has prompted the application of established risk-analysis techniques to improve the design safety assessment process. In the United States, Title 46 of the Code of Federal Regulations represents the regulatory requirements applicable to the design, construction, and operation of U.S.-flagged ships; it mandates the use of a qualitative failure analysis technique. In particular, Part 62, Vital System Automation, presents the minimum requirements for vessels' automated systems. One failure analysis technique that has been applied in both national and international marine regulations is Failure Mode and Effects Analysis (FMEA), which assumes a failure mode occurs in a system/component through a certain failure mechanism; the effect of this failure is then evaluated.

Additional hazards exist for high-speed craft. The International Code of Safety for High-Speed Craft (HSC), adopted by the International Maritime Organization (1995), provides regulations for high-speed (low-displacement) craft. The code's safety management philosophy is based on risk assessment and management. FMEA is a required part of HSC Code compliance, mandating provision of failure analysis to assist in safety management. The FMEA procedure is specified as an appendix to the HSC Code.

The International Organization for Standardization issued the ISO 9000 series of quality management standards in 1988. The series required organizations to develop formalized Quality Management Systems focused on the needs, wants, and expectations of customers.QS-9000, the automotive equivalent to ISO 9000, was developed by a task force representing Chrysler Corporation, Ford Motor Company, and General Motors Corporation; it is a quality system standard that focuses on helping automotive suppliers satisfy customers' requirements. QS-9000 is being replaced by a newer, related standard called ISO/TS 16949, which includes all of ISO 9000, QS-9000, and many European standards. In accordance with ISO/TS 16949 standards, compliant automotive suppliers use Advanced Product Quality Planning (APQP), including design and process FMEAs, and develop control plans.

In the automotive industry, APQP standards provide a process-oriented method of identifying and establishing the steps necessary to ensure customer satisfaction. As a critical element for implementing the APQP process, a control plan provides "a structured approach for the design, selection, and implementation of value-added control methods for the total system" (Chrysler Corporation, Ford Motor Company, and General Motors Corporation, 1994). Control plans aid the in manufacture of quality products according to customer requirements in conjunction with ISO/TS 16949. An emphasis is placed on product- and process-variation minimization, which is also the focus of the Six Sigma methodologies. ISO/TS 16949–compliant automotive suppliers must use FMEAs in the APQP process and in the development of their control plans to prevent failures when designing and making new products.

As a step to standardize Failure Mode and Effect Analysis efforts, the Automotive Industry Action Group (AIAG) and the American Society for Quality (ASQ) copyrighted industrywide FMEA standards in February 1993. It is the technical equivalent of the Society of Automotive Engineers' procedure SAE J-1739, which was jointly developed by Chrysler, Ford, and General Motors under the sponsorship of the U.S. Council for Automotive Research (USCAR). The standards are presented in an FMEA manual entitled "Potential Failure Modes and Effects Analysis in

Design (Design FMEA) and Potential Failure Modes and Effects Analysis in Manufacturing and Assembly Processes (Process FMEA)." The manual provides general guidelines for developing a plan to use FMEA for examining at all of the ways a product or process can fail, analyze risks, and take action where warranted.

The design and manufacture of safe medical devices inherently includes risk management as a key objective. The FDA Quality System Regulation includes risk analysis "where appropriate" as part of design validation in the Design Control requirements. The European CE Marking regulation, and the Medical Devices Directive in its Essential Requirements, requires manufacturers "taking account of the generally acknowledged state-of-the-art" to "eliminate or reduce risks as far as possible." An International Electrotechnical Commission publication (IEC 812, 1997) describes Failure Mode and Effects Analysis (FMEA) and Failure Mode, Effects and Criticality Analysis (FMECA); it gives guidance as to how objectives discussed in the following sections can be achieved when using FMEA and FMECA as risk-analysis tools.

As described here, the application of a Failure Mode and Effects Analysis has been increasing in various industries. Its essential function of identifying the potential failure modes within an industrial system is an integral part of engineering robust products with Six Sigma. While revealing to engineers how a system may fail, FMEA provides the foundation for failure prevention and risk management.

5.2 Using FMEA to Prevent Failure Before Any Harm Is Done

The Failure Mode and Effect Analysis method can be used to analyze potential reliability problems early in the development cycle when it is cost-effective to take corrective actions to prevent any issues—that is, improving reliability through design. Customers are placing increased demands on companies for high-quality, reliable products. Then too, the increasing capabilities and functionality of many products are making it more difficult for manufacturers to maintain quality and reliability. Traditionally, quality and reliability have been achieved through extensive prerelease testing and use of techniques, such as probabilistic reliability modeling, that are performed during the late stages of development. The challenge is to design in quality and reliability as early as possible.

As a systematic methodology for identifying potential failure of a product or service and then determining the frequency and impact of that failure, FMEA is very useful. It can also identify failure modes and their effect on the operation of a product and what actions are necessary to mitigate failures. Anticipating what might go wrong with a product is a crucial step. Although preparing for every failure is not possible, the development team should formulate as extensive a list of potential failure modes as possible.

When engineering for robust design with Six Sigma, FMEA is a *proactive* product development tool to identify and prevent design and/or process errors before they occur. The Failure Mode and Effects Analysis can also be used to rank and to prioritize possible causes of failure, as well as to develop and implement preventive actions, with responsible persons assigned to complete them. The early and consistent use of FMEAs during the design process allows Six Sigma teams to design out failures and produce reliable, safe, and customer-pleasing products. FMEAs can capture historical information for use in product improvement too.

5.3 Identifying Functions and Failure Modes

The FMEA process starts with identifying functions. A *function* is an intended purpose of a product or process being analyzed. If a system is being considered, functions of individual subsystems should also be identified. Potential failure modes, or categories of failure, can then be identified by describing the way in which the object fails.

Failure mode, sometimes described as categories of failure, refers to the different ways that something can fail to provide what was anticipated. That is, it details the way in which a product or process could fail to perform its function (design intent or performance requirement) as described by the needs, wants, and expectations of internal and/or external customers. The *customer* could be the next operation, subsequent operations, installation and service personnel, or the end user. Failure modes fall into one of five possible categories, as follows:

- Complete failure
- Partial failure
- Intermittent failure
- Failure over time
- Overperformance of a function

Example 5.1

Suppose that "Provide light at 180 ± 0.8 candelas" is the defined function of a lightbulb circuit. All of the units for measuring and defining light are based on the *candela,* which is the luminous intensity from a small source in a particular direction. This unit was originally based on the light emitted from a candle's flame.

The following failure modes could be identified:

- No light
- Dim light

- Erratic blinking light
- Gradual dimming of light
- Too bright a light

The purpose of these five failure mode groupings is to assist the FMEA team with the identification of all possible failure modes; looking for these groupings may reveal some unusual failure possibilities that otherwise would not have been considered. Poorly defined functions also may be revealed.

In the example in Figure 5-1, a light that does not turn off (overperforms its function) is a product failure, even though the function—"Provide light at 150 ± 0.8 candelas"—is not a failure. This implies the need for an additional function, such as "Default to off when not in use," which may have been overlooked when functions were originally identified. The original function can be rephrased as: "Provide light at 150 ± 0.8 candelas when on." A partial, intermittent, gradual, or overperformance type of failure of one function may be a complete failure of another unidentified function. Use of the failure mode categories can help reveal such functions.

During the progression of time, a failure mode comes between a cause and an effect. One of the most confusing issues for new practitioners of FMEA is that any *cause* that itself has a cause might be a failure mode. Any *effect* that itself has an effect might also be a failure mode. In different contexts, a single event may be a cause, an effect, and a failure mode.

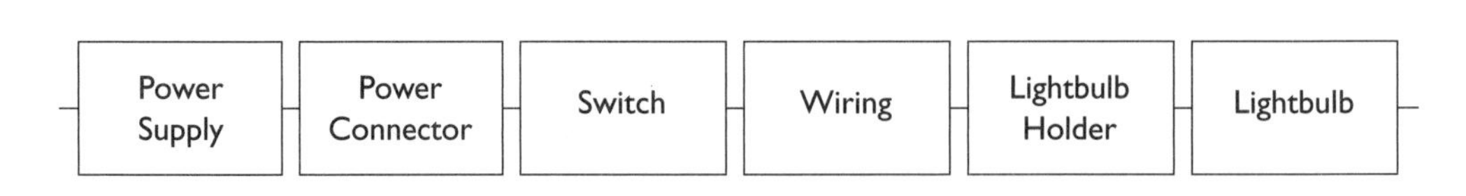

Figure 5-1 A FMEA often starts with a functional diagram like this one for Thomas Edison's lightbulb.

For example, an analysis of the glass cover of a lightbulb that reveals "Allows Excessive Moisture" would be a failure mode. Besides maintaining a vacuum within the lightbulb, one of the design functions of the glass cover is to prevent exposure of the filaments to excessive moisture during normal operation. *Allows Excessive Moisture* is a failure mode because one CTQ for the glass cover is to prevent moisture from getting in. Causes—Lightbulb Design Defect and Environmental Exposure—appear before Failure Mode in Figure 5-2.

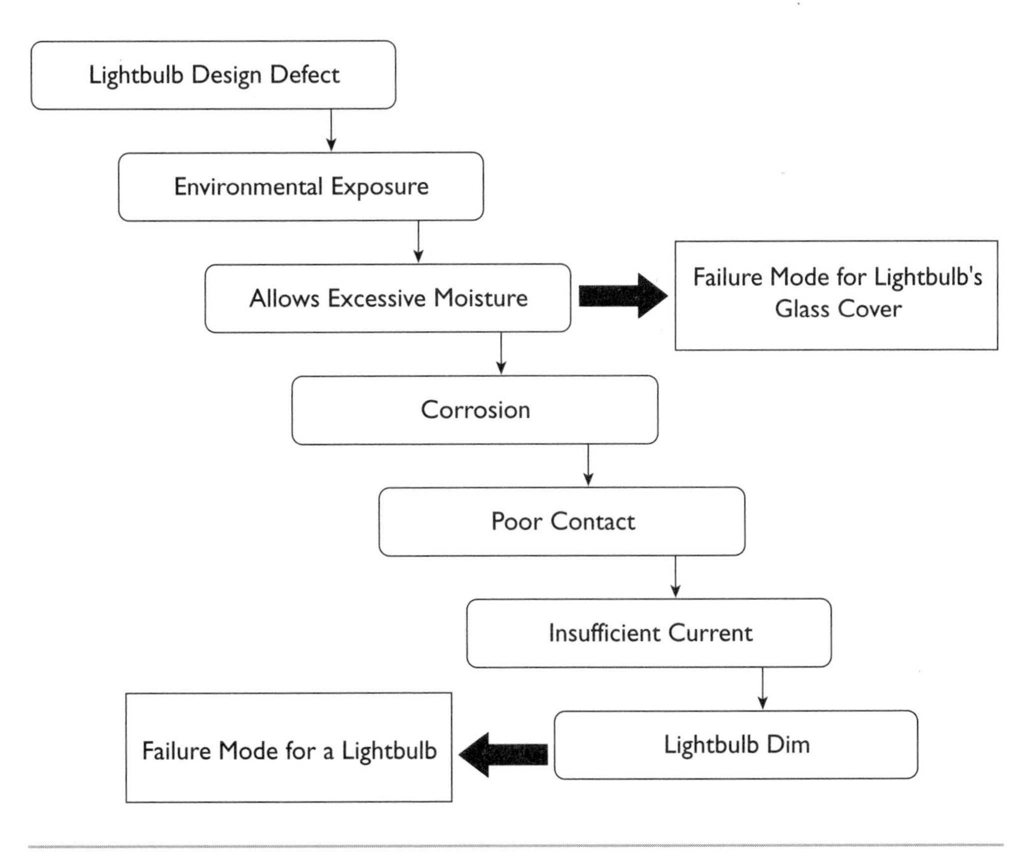

Figure 5-2 Events that could occur during the life of a lightbulb.

Effects—Corrosion, Poor Contact, Insufficient Current, and Lightbulb Dim—appear after the glass cover's failure mode in the figure. However, in the analysis of the lightbulb, a different function and failure mode must be considered. The lightbulb is designed to provide light of specific intensity when the device is activated during its expected lifetime. This is one of its functions, or design CTQs. As shown in Figure 5-2, a dim bulb is a failure to provide the specified intensity of light and is therefore a failure mode of the lightbulb.

This example illustrates that causes, effects, and failure modes can change depending on the function being analyzed. Functions change depending on the object of the analysis, either product or process. Therefore, an early, important step in a FMEA is to clearly define the scope—the component, system, or product—that is to be analyzed.

Most real-world systems do not follow a simple cause–effect model. A single cause may have multiple effects. A combination of causes may lead to an effect, or they may lead to multiple effects. Causes can themselves have causes, and effects can have subsequent downstream effects. The failure mode must also be considered in all of these instances.

Figure 5-3 illustrates the fishbone[1] relationship among a function, a failure mode, and potential causes and effects. In the FMEA model that is presented here, causes do not automatically result in a failure mode. The term *potential* is used to describe causes—to indicate this uncertainty. The model also assumes that all effects will result given that the failure mode has occurred. Therefore, "potential" is not used to describe effects.

1. A fishbone diagram is a Six Sigma tool used to solve critical quality problems by brainstorming causes and effects and logically organizing them by branches.

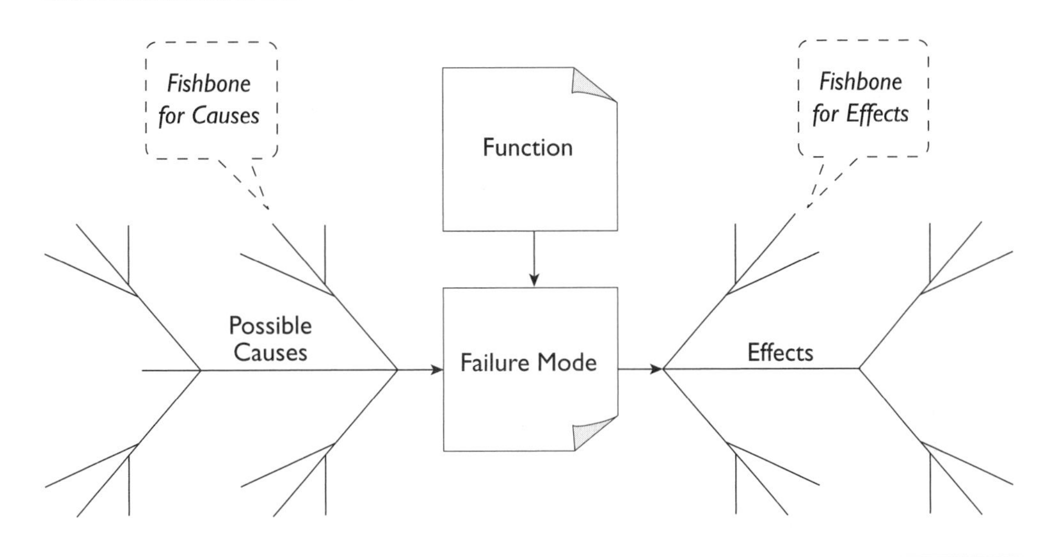

Figure 5-3 Diagram of the fishbone relationship among function, failure mode, causes, and effects.

5.4 IDENTIFYING EFFECTS AND SEVERITY

The Failure Mode and Effect Analysis is an integral part of the Six Sigma process for developing robust products. FMEA's elements are the building blocks of related information that comprise an analysis, and a team approach is almost always essential to identify the elements.

Although actual document preparation is often the responsibility of an individual, FMEA input should come from a multidisciplinary team. Such a team should consist of knowledgeable individuals with expertise in design, manufacturing, assembly, service, quality, and reliability. The responsible engineer typically leads the FMEA team. Members and leadership may vary as system, product, and process designs mature.

Failure modes often result in a departure of a Critical-to-Quality (CTQ) characteristic from its intended level, or state, that occurs with severity sufficient to cause an associated product or service not to meet a specified requirement. After functions and failure modes have been established, the next step in the FMEA process is to identify potential downstream consequences when a failure occurs. This should be a team brainstorming activity. After consequences have been identified, they must be added to the FMEA model as effects.

5.4.1 EFFECTS AND CONSEQUENCES

Generally, it is assumed that effects always occur when the failure mode occurs; there is no representation for the likelihood that a failure mode will result in an effect. The procedure for identifying potential consequences is applied to account for unlikely or remote consequences. The procedure explicitly associates effects with the circumstances under which they occur through the identification of additional failure modes. The following are the procedure's steps:

1. Begin with a failure mode (referred to as FM-1) and a list of all of its potential consequences.
2. Separate the consequences that can be assumed to result whenever FM-1 occurs; identify these as the effects of FM-1.
3. Write additional failure modes for the remaining consequences—ones that could result when FM-1 occurs, depending on the circumstances under which it occurred. The new failure modes imply that unlikely consequences will result by including the circumstances under which they occur.
4. Separate the consequences that can be assumed to result whenever additional failure modes and their special circumstances occur; identify these consequences as effects of the additional failure modes.

The example that follows illustrates the procedure for identifying potential consequences. During the effects brainstorming session, the team tends to identify very severe consequences and the unlikely circumstances under which they may occur. When analyzing the lightbulb, a team member may observe that the bulb could prematurely burn out when being used as a flashlight, the resulting darkness causing the user to trip, fall, and be injured. Another member may observe that atmospheric pressure variation could cause the bulb to explode while being used for an eye examination, resulting in injury. Such extraneous predicaments are typical of brainstorming and are to be expected. But, rather than write a new failure mode for every bizarre situation recorded by the team, the events should be grouped into a broad effect category, such as "injury or death."

Ultimately, effects need to be categorized into one of ten groups, according to severity. It is advantageous to write failure modes that encompass all of the effects in a severity grouping (e.g., "Fails to provide 150 ± 0.8 candela of light under critical conditions"). All product failures that lead to injury or death are automatically included; there is no need to attempt to identify all of the circumstances under which injury or death could result. The tradeoff for this convenience is that the likelihood of failure under any critical situation must be factored into an occurrence rating.

Now that the unlikely consequences of injury or death are represented by their own failure mode, the effects will be analyzed separately, independent of the original failure mode—"Fails to provide 150 ± 0.8 candela of light when on." The original failure mode should be modified to read: "Fails to provide 150 ± 0.8 candela of light under nominal conditions," which implies that effects may be more severe under different circumstances. The procedure for identifying potential consequences is used in combination with a modified definition of occurrence to represent effects that do not always result when the cause of failure occurs. The procedure for identifying potential consequences and the occurrence redefinition are presented as an improvement over cause–effect relationships in existing FMEA models.

5.4.2 RANKING EFFECTS' SEVERITY

The first step in analyzing risk is to quantify the severity of the effects. Effects are rated on a 1 to 10 scale, 10 being the most severe. The team should agree on consistent evaluation criteria and a sensible ranking system. The design and process ranking system presented in Table 5-1 is based on AIAG standards. Effects are evaluated as a group when assessing risk, even though they are assigned severity values individually.

Table 5-1 A Ranking System for the Severity of Effects for a Design FMEA

Effect	Description of Severity	Rank
Hazardous, without warning	Failure affects safe product operation or involves noncompliance with government regulation without warning	10
Hazardous, with warning	Failure affects safe product operation or involves noncompliance with government regulation with warning	9
Very high	Product is inoperable with loss of primary function	8
High	Product is operable, but at a reduced level of performance	7
Moderate	Product is operable, but comfort or convenience items are inoperable	6
Low	Product is operable, but comfort or convenience items operate at a reduced level of performance	5
Very low	Fit and finish or squeak and rattle—items do not conform; most customers notice defect	4
Minor	Fit and finish or squeak and rattle—items do not conform; average customers notice defect	3
Very minor	Fit and finish or squeak and rattle—items do not conform; discriminating customers notice defect	2
None	No effect	1

It is assumed that all effects will result if the failure mode occurs. Therefore, the most serious effect takes precedence when evaluating risk potential. This model accounts for causes that have multiple effects. Issuing design and process changes can reduce severity ratings.

Corrective actions often result in lower severity, less occurrences, or a better detection rating (see Section 5.6). Design or process revision may result in lower severity and occurrence ratings. Adding validation or verification controls can reduce the need for detection. The revised ratings need to be documented against the originals on the tabular FMEA form. If no action is recommended, the decision not to act should also be noted. Effective follow-up programs are also necessary because the purpose of the Failure Mode and Effect Analysis process is defeated if any recommended actions are left unaddressed.

5.5 Understanding Causes

Cause is a factor (X) that has an impact on a response variable (Y); it is a source of variation in a product or a system. In an FMEA, cause is an identified reason for the presence of a failure mode or problem. A cause is the original reason for a certain product's failure mode. When causes are prevented or removed, the failure mode will be avoided or eliminated.

After effects and severity have been addressed, the next step is to identify causes of failure modes. This is another team activity. Identification should start with failure modes that have the most severe effects. In a Design FMEA, design deficiencies that result in a failure mode are causes of failure. Design deficiencies that induce a manufacturing or assembly error are also included in Design FMEAs as causes. Table 5-2 shows a ranking system for the occurrence of failure in a Design Failure Mode and Effect Analysis.

Table 5-2 A Ranking System for the Occurrence of Failure in a Design FMEA

Probability	Rate	Rank
Very high: Failure almost inevitable	> 1 in 2	10
Hazardous, with warning	1 in 3	9
High: Repeated failures	1 in 8	8
	1 in 20	7
Moderate: Occasional failures	1 in 80	6
Low to very low	1 in 400	5
	1 in 2000	4
Low: Relatively few failures	1 in 15,000	3
Very minor failures	1 in 150,000	2
Remote: Failure unlikely	< 1 in 1,500,000	1

In a Process Failure Mode and Effect Analysis (Process FMEA), causes are specific errors described in terms of something that can be corrected or controlled. The Process FMEA assumes that the product is adequately engineered and that it will not fail because of a design deficiency. This does not imply that all inputs to the process meet engineering specifications. Variation in purchased parts and materials used in the process should be considered in the Process FMEA. *Special-cause* variation is a shift in output caused by a specific factor such as environmental conditions or process input parameters. It can be accounted for directly and potentially removed, and such a cause variation is a measure of process control for product development.

5.6 Assessing Current Controls

Current controls (design and process) are the mechanisms that prevent the cause of the failure mode from occurring or that detect the failure before it reaches the customer. Design and process controls are grouped according to their purpose, as follows:

- *Type 1* controls prevent the cause or failure mode from occurring or reduce its rate of occurrence.
- *Type 2* controls detect the cause of the failure mode and lead to corrective action.
- *Type 3* controls detect the failure mode before the product reaches the customer—that is, the next operation, subsequent operations, or the end user.

The distinction between controls that prevent failure (Type 1) and controls that detect failure (Types 2 and 3) is important. Type 1 controls reduce the likelihood that a cause or failure mode will occur and therefore affect occurrence ratings. Type 2 and Type 3 controls detect causes and failure modes, respectively, and therefore affect detection ratings.

Detection values are associated with current controls. Detection is a measurement of the ability of Type 2 controls to detect causes or mechanisms of failure or the ability of Type 3 controls to detect subsequent failure modes.

Table 5-3 details a design and process ranking system for the detection of a cause of failure or failure mode in a Design FMEA; it is based on AIAG standards. One detection value is assigned to the system of current controls, which represents a collective ability to detect causes or failure modes.

Table 5-3 A Ranking System for the Detection of a Cause of Failure or Failure Mode in a Design FMEA

Detection	Description of Likelihood	Rank
Absolutely uncertain	Uncertain that a design control will detect a potential cause of failure or subsequent failure mode; or there is no design control	10
Very remote	Very remote chance that a design control will detect a potential cause of failure or subsequent failure mode	9
Remote	Remote chance that a design control will detect a potential cause of failure or subsequent failure mode	8
Very low	Very low chance that a design control will detect a potential cause of failure or subsequent failure mode	7
Low	Low chance that a design control will detect a potential cause of failure or subsequent failure mode	6
Moderate	Moderate chance that a design control will detect a potential cause of failure or subsequent failure mode	5
Moderately high	Moderately high chance that a design control will detect a potential cause of failure or subsequent failure mode	4
High	High chance that a design control will detect a potential cause of failure or subsequent failure mode	3
Very high	Very high chance that a design control will detect a potential cause of failure or subsequent failure mode	2
Almost certain	Almost certain that a design control will detect a potential cause of failure or subsequent failure mode	1

Controls can be grouped and treated as a system when they operate independently because each individual control increases overall detection capabilities. The intent of a process control plan is to control product

CTQs and the associated process variables to ensure capability (around the identified target or nominal) and stability of the product over time. Through a Design Failure Mode and Effect Analysis, the risks associated with something potentially going wrong (creating a failure mode) within the product being designed can be identified. Based on the FMEA, it is possible to determine what controls need to be placed in ongoing process to catch and remove any failure modes at various stages of product development.

5.7 Using FMEA for Risk Management

A design *risk assessment* is the act of determining potential risk during a design process, either in a concept design or a detailed design. It provides a broader evaluation of designs beyond just CTQs and an assessment enables you to prevent possible failures and/or to reduce the impact of potential failures. It ensures a rigorous, systematic examination of the reliability of the design and allows you to capture system-level risk.

The fundamental purpose of the FMEA is to recommend and to take actions that reduce risk. A Design Failure Mode and Effect Analysis would be a quantified estimate of the criticality of CTQ internal failures and Voice-of-Customer (VOC) performance factors using the classic FMEA form, but for design reviews and technology transfers. Performance that satisfies customers' needs use VOCs and design parameters to help meet requirements; also, the manufacturability requirements might be called CTQs. Both would be appraised in the Design FMEA document and then updated as the product moves into production based on real failure frequencies and severities, becoming a living risk-management and design feedback document.

The systematic, proactive FMEA method of evaluating a process identifies the opportunities for failure, or failure modes, in each step. Every failure mode gets a numeric score that quantifies (1) the likelihood that the failure will occur (O), (2) the likelihood that failure will be detected (D), and (3) the amount of harm or damage (i.e., severity, S) a failure mode may cause to a person or to equipment. The product of these three scores is the *Risk Priority Number* (RPN)—a numeric assessment of risk assigned to a process or steps in a process—for that failure mode. The equation form is RPN = S • O • D. The RPNs' sum for the failure modes is the overall Risk Priority Number for the process.

As an organization works to improve a process, product, or system, it can anticipate and compare the effects of proposed changes by calculating hypothetical RPNs of different scenarios. Just remember that the RPN is a measure for comparison within one process only; it is not a measure for comparing risk between processes, products, or organizations.

For the failure mode—Allows Excessive Moisture—in the lightbulb example, severity is 6, occurrence is 5, and detection is 4 (see Table 5-4). Thus, RPN = S • O • D = 120. In FMEA, CTQs are primarily product characteristics for which reasonably anticipated variation could significantly affect a product's safety or compliance with governmental standards or regulations or is likely to significantly affect customer satisfaction with a product. Here, identifying CTQs is a process to find out about potential failures of components, systems, and products and their effects on equipment and human safety. This process helps introduce equipment safety features beforehand to avoid potential failures and possible consequent accidents.

Using FMEA for risk management, CTQs predominantly reflect product or process requirements that affect compliance with government regulation or safe product function and that require special corrective actions or control plans. In a FMEA, any characteristic with a severity rating of 9 or 10 is a CTQ, which requires a special control to ensure detection. Examples of product or process requirements that could be CTQs include

Table 5-4 FMEA Form for a Lightbulb

Part Name / Part Number	Function	Failure Mode	Causes of Failure	Effects of Failure	Current Control	R = S • O • D				Recommended Corrective Action	Action Taken
						S	O	D	R		
Penlight case	Protect internal parts from excess moisture	Allows excess moisture	Case wear due to design's sensitivity to environmental exposure	Corrosion, poor contact, insufficient current, dim bulb	Case material selection	6	5	4	120	Reliability testing	Testing performed on June 14

dimensions, weights, specifications, tests, assembly sequences, tooling, joints, torque, welds, attachments, storage, shipping, installations, product usage, and so on. Special actions or controls necessary to meet these requirements may involve design, test, manufacturing, assembly, a supplier, a shipper, monitoring, inspection, or something else.

CTQs require special controls because they are important to customer satisfaction. Besides the safety- and compliance-related CTQs described before, severity ratings between 5 and 8, coupled with an occurrence rating greater than 3, also designate Critical-to-Quality characteristics that require special design controls to ensure detection of potential failure modes.

5.8 SUMMARY

A Failure Mode and Effect Analysis is a systematic method to ensure that every conceivable failure of a design and/or process has been considered. Doing an FMEA is a simple analysis technique that identifies the ways in which something might fail while in service and the effects that would be experienced by customers. Potential causes of failure can be explored and the design controls that are normally applied can be evaluated for their effectiveness. By producing a numerical rating, it is easy to identify and to prioritize problem areas in order of importance. Corrective actions can then be targeted at the high-risk issues and another analysis can determine the reduction in risk that is achieved.

A Design Failure Mode and Effect Analysis captures the relationship between customer requirements, how a product can fail to meet the requirements, the effects of the failures, problems with the design that cause failures, and how the design will be validated to prove it will not fail. Rating columns exist for the effects of failure, the probability of failure, and the effectiveness of design validation. These ratings are multiplied together to achieve a Risk Priority Number for each product failure and cause combination. Rating numbers typically range from 1 to 5 or 1 to 10 for each of the columns, with higher numbers designating unacceptable conditions.

A Design FMEA can also determine potential risk in a design process, either in a concept or a detailed design. It provides an evaluation of product design beyond just CTQs and enables a Six Sigma team to eliminate possible failures and/or to reduce the impact of potential failures. Doing a Design FMEA ensures a rigorous, systematic examination of the reliability of the design and allows engineers to capture system-level risk.

BIBLIOGRAPHY

Andrews, J. D., and T. R. Moss. *Reliability and Risk Assessment, First Edition.* Essex, UK: Longman Group, 1993.

Aven, Terje. *Reliability and Risk Analysis, First Edition.* Essex, UK: Elsevier Applied Science, 1992.

Bell, Daniel, Lisa Cox, Steve Jackson, and Phil Schaefer. "Using Causal Reasoning for Automated Failure and Effects Analysis (FMEA)." Proceedings of the Annual Reliability and Maintainability Symposium (pp. 343-53), Las Vegas, 1992.

Bouti, Abdelkader, and Daoud Ait Kadi. "A State-of-the-Art Review of FMEA/FMECA." *International Journal of Reliability, Quality and Safety Engineering,* 1(4): 515–43, 1994.

Chrysler Corporation, Ford Motor Company, and General Motors Corporation. "Advanced Product Quality Planning and Control Plan," 1994.

Henley, Ernest J., and Hiromitsu Kumamoto. *Probabilistic Risk Assessment, First Edition.* Piscataway, NJ: IEEE Press, 1992.

International Electrotechnical Commission. *Procedure for Failure Mode and Effects Analysis (FMEA)*, IEC 812, 1997.

International Maritime Organization. *International Code of Safety for High-Speed Craft* (pp. 175–85), 1995.

Kara-Zaitri, Chakib, Alfred Z. Keller, Imre Barody, and Paulo V. Fleming. "An Improved FMEA Methodology." Proceedings of the Annual Reliability and Maintainability Symposium (pp. 248–52), Orlando, FL, 1991.

Kara-Zaitri, Chakib, Alfred Z. Keller, and Paulo V. Fleming. "A Smart Failure Mode and Effect Analysis Package." Proceedings of the Annual Reliability and Maintainability Symposium (pp. 414–21), Las Vegas, 1992.

The Office of the Federal Register National Archives and Records Administration, Title 46—Code of Federal Regulations, Part 62—Vital System Automation (Revised), October 1997.

Pelaez, C. Enrique, and John B. Bowles. "Applying Fuzzy Cognitive-Maps Knowledge-Representation to Failure Modes and Effects Analysis." Proceedings of the Annual Reliability and Maintainability Symposium (pp. 450–59), Washington, DC, 1995.

Price, C. J., J. E. Hunt, M. H. Lee, and R. T. Ormsby. "A Model-Based Approach to the Automation of Failure Mode Effects Analysis for Design." Proceedings of the Institution of Mechanical Engineers (ImechE), Part D. *The Journal of Automobile Engineering*, 206: 285–91, 1992.

Price, C. J., David R. Pugh, Myra S. Wilson, and Neal Snooke. "The Flame System: Automating Electrical Failure Mode and Effects Analysis (FMEA)." Proceedings of the Annual Reliability and Maintainability Symposium (pp. 90–95), Washington, DC, 1995.

Roland, Harold E., and Brian Moriaty. *System Safety Engineering and Management, Second Edition*. New York: John Wiley & Sons, 1990.

Russomano, David J., Ronald D. Bonnell, and John B. Bowles. "A Blackboard Model of an Expert System for Failure Mode and Effects Analysis." Proceedings of the Annual Reliabilty and Maintainability
Las Vegas, 1992.

nd *Effect Analysis—FMEA from Theory*
SQC Quality Press, 1995.

hiainen. *Quality Management of Safety*
: Elsevier Science Publishers, 1993.

nkar, Farhad Moghadam, and J. Garcia-
the Application of Process Simulations
es." Proceedings of the IEEE/SEMI
Manufacturing Conference (pp. 515–43),

Roush. *What Every Engineer Should Know*
Management. New York: Marcel Dekker,

ux. "Failure Mode and Effects Analysis
nt in a Semiconductor Manufacturing
of the IEEE/SEMI Advanced Semicon-
ference (pp. 215–32), Essex Junction, VT,
1994.

6 The P-Diagram

Laying Out a Robust Design Strategy

To layout a robust design strategy, the product development team needs to identify the inputs and outputs associated with the design concept. Besides these, the team should understand which parameters can be controlled cost-effectively and which cannot be controlled. Those that can be specified are called control factors, and parameters beyond the control of the designer are called noise factors. A large number of product failures and the resulting Cost of Poor Quality (COPQ) are a result of neglecting noise factors during the early design stages. Robust design involves deciding the best values and/or levels for the control factors in the presence of noise factors. The Parameter Diagram (P-Diagram) is an essential tool for every robust design project.

6.1 Experimental Design and the P-Diagram

In an experiment's design, one or more design variables is deliberately changed in order to observe the effect that change has on one or more response variables for product design. Experimental design is an efficient procedure for planning experiments so that the data obtained can be analyzed to yield valid and objective information about the engineered product.

Experimental design begins with determining the objectives of an experiment and selecting the variables to study. Through the design, a detailed plan can be laid out in advance of doing the experiment. Well-planned experimental designs maximize the amount of information that can be obtained for a given amount of effort. The statistical theory underlying experimental design generally begins with the P-Diagram concept.

6.1.1 P-Diagram

The Parameter Diagram, or P-Diagram, is a useful tool in brainstorming and documenting control and noise factors.

- *Control factors*—parameters that can be specified by designers
- *Noise factors*—parameters that are beyond the control of designers

A P-Diagram is essentially a schematic drawing that encompasses a control factor, a noise factor, a signal factor, and a response variable. In Figure 6-1, the signal factor represents the inputs to the system to produce the intended response.

A control factor is a design parameter for which a nominal value can be chosen as desired to reduce the sensitivity of performance against noises. The goal of robust design is to reduce variation in performance, or the response, in the presence of noise. This is achieved by selecting an optimal setting of nominal values for some of the design variables such that performance is less sensitive to noise variability and input signals. A robust design should optimize the relationship between the input signal and the output signal. Ideally, this relationship should be linear and

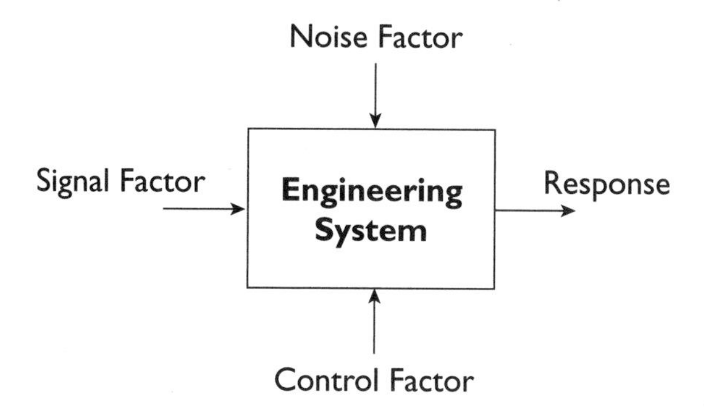

Figure 6-1 A P-Diagram for laying out robust designs.

tuned such that the transfer function is stable and consistent across the range of customer noises that a system will be subjected to.

6.1.2 Control Factors

Control factors are the factors whose nominal values can be adjusted, or the design variable that can be specified, ideally with minimal impact on cost, resource, and technology considerations. The objective of robust design experiments is to identify control factor settings at which the influence of noise is at a minimum. Control factors need to be specified by the product development team. For example, designing a new car brake involves the following control factors:

- Pad area
- Pad material
- Rotor thickness
- Caliper type

6.1.3 NOISE FACTORS

Noise factors are parameters that influence system, subsystem, and/or component variability and that are difficult, expensive, or impossible to control; in other words, uncontrolled or difficult-to-control factors. The product development team needs to make responses insensitive to noise. The following are a few important noise factor types to consider when engineering a general system.

Customer-usage profiles: The environment in which a product and/or process works and the load to which it is likely to be subjected. This includes ambient temperature, humidity, vibration, and operator usage.

Manufacturing unit-to-unit variation: Inevitable variation in a manufacturing process leads to variation in the product from unit to unit. Examples include differences in torque on a fastener, differences in diameter from machining, and differences in raw materials.

Deterioration: Individual components may deteriorate causing a decrease in product and/or process performance as time passes. Product deterioration is a result of wear over time.

6.1.4 SIGNAL FACTORS

Signals (or inputs) are factors that are set by the user of the product or process to achieve the intended function for the response. It is a translation of customers' intent to measurable engineering terms with a defined operating range. The following are a few examples for signal factors for car brakes:

- Force on the brake pedal
- Force on the accelerator

- Temperature setting of a car's heater
- Steering wheel angle
- Turning diameter (angle of turn)

6.1.5 RESPONSE VARIABLES

The response (or output) variables are the translation of the customer's perceived requirement into measurable engineering terms. The starting point for an engineer who needs to create a robust design for a product is identifying the Critical-to-Quality characteristics (CTQs). Once identified, he or she can focus on making these characteristics robust so that the product responds to the customer in the desired way. Figure 6-2 identifies noise factors that impact stopping distance and therefore should be addressed by making the example braking system more robust.

For a car braking system, the stopping distance is the response that is critical to its quality. The response, stopping distance, should be insensitive to road conditions, vehicle load, ambient temperature, mechanical wear over time, and vehicle speed.

> ➢ Going fast requires that one can stop fast too
> ➢ Brakes work with a friction pad
> ➢ Friction converts car momentum into heat
> ➢ Heat is dissipated over the brake disk
> ➢ Stopping distance depends on mechanical, thermal, and control designs

Figure 6-2 Stopping distance—a CTQ characteristic for a braking system.

6.1.6 IDEAL FUNCTION

When the engineer attempts to translate the customer's requirements into measurable engineering terms, they are modeled using mathematical functions. The functions are given in terms of equations that express the laws of physics, which govern the relationships the engineer is modeling in the system.

The model created describes how the system should perform in an ideal world and is referred to as the *ideal function;* that is, how the system, the car's braking mechanism, would perform if everything in the higher-level system were to function perfectly as designed. The equations show the mathematical input–output relationship between the output (response) and the input (signal)variables. The ideal function can be in a linear or nonlinear form with or without a constant term (called *intercept*).

The P-Diagram is based on the concept of converting 100 percent of the input energy (input signal) into 100 percent of the ideal function. Any engineered system reaches its ideal function when all of its applied energy (input) is transformed efficiently into creating desired output energy.

In reality, nothing functions like this. Every system is less than 100 percent efficient in its energy transformation; the resulting loss goes toward creating unintended functions, or error states. A signal-to-noise ratio, often written S/N or SNR, expresses the ratio between the ideal function and the error state.

To produce a robust design, one constructs an experimental design for control factors using a factorial design (inner array) and an outer array for noise factors. Then physical experiments are carried out according to the settings of the inner and outer arrays with the response recorded under the outer array. For each setting of control factors, one calculates the sample mean and variance, hence, the SNR. A P-Diagram for a car's brakes is shown in Figure 6-3.

Another example: The arch enemy of picture clarity on any monitor is noise—electronic noise, which is present to some extent in all video signals. Noise manifests itself as snow or graininess over the whole picture on the monitor. There are several sources of noise, including poor circuit

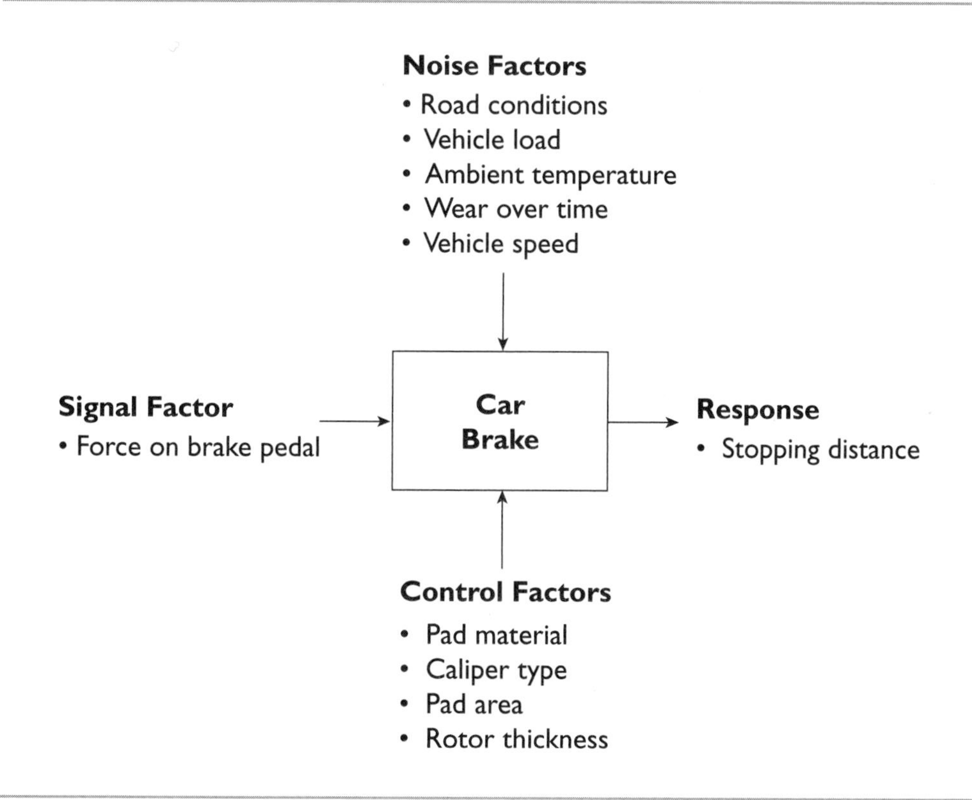

Figure 6-3 A P-Diagram for a car's brakes.

design, heat, overamplification, external influences, automatic gain control, and transmission systems (e.g., microwave, infrared, etc.). The important factor that determines tolerance to noise is the amount of it in the video signal—the S/N ratio. Note that every time a video signal is processed in any way, noise is introduced.

In robust design, the signal-to-noise ratio is an engineering term for the ratio between the magnitude of a signal (meaningful information) and the magnitude of background noise. Because many signals have a very wide dynamic range, SNRs are often expressed in terms of the logarithmic decibel (dB) scale. As shown in Figure 6-4, ideal function is a description of how the engineered system should perform its intended function perfectly.

For example, communications engineers always strive to maximize the S/N ratio. Traditionally, this has been done by using the narrowest possible receiving-system bandwidth consistent with the data speed desired; however, there are other methods. In some cases, spread-spectrum tech-

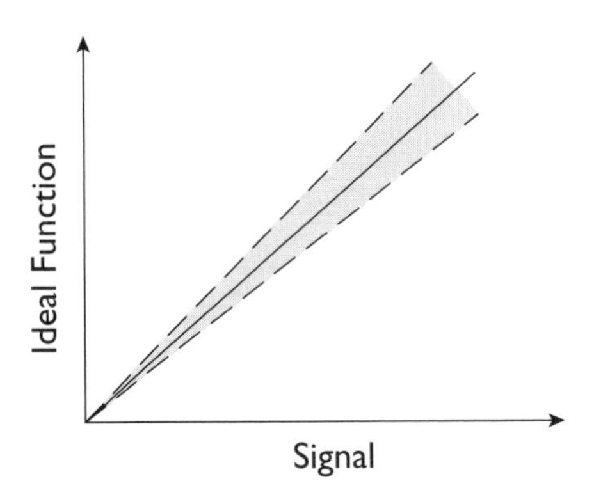

Figure 6-4 Ideal function and S/N ratio.

niques can improve system performance. The S/N ratio can be increased by providing the source with a higher level of signal output power if necessary. In some high-level systems, such as radio telescopes, internal noise is minimized by lowering the temperature of the receiving circuitry to near absolute zero (–273°C or –459°F). In wireless systems, it is always important to optimize the performance of the transmitting and receiving antennas.

6.2 QUALITY LOSS FUNCTION

Quality loss function (QLF) is a process used to measure quality. It establishes a financial measure of customer dissatisfaction with a product's performance as it deviates from a target value. Both average performance and variation are critical measures of quality. Engineering robust designs that are insensitive to uncontrolled sources of variation improves quality. In Six Sigma, Taguchi's Quality Loss Function method is used to approximate the financial loss for any particular deviation in a product using the best specification target for the amount of variation in the manufacturing process (Taguchi and Yokoyama, 1993). The greater the deviation from the target, the greater is the quality loss. To fully understand the Taguchi's Quality Loss Function, this section starts with a review of the traditional QLF.

6.2.1 TRADITIONAL QUALITY LOSS FUNCTION

The most common approach to quality is often referred to as a *goalpost* mentality of quality. Traditional thinking deems anything within specification limits as "OK" and anything outside the limits as "not OK." The usual representation of a quality loss function, which relates to the tolerance range of the product's expected performance, would look like Figure 6-5—a standard quality graph. Its *x*-axis represents Performance of CTQ, while the *y*-axis represents a Loss to Society (i.e., customer dissatisfaction). The

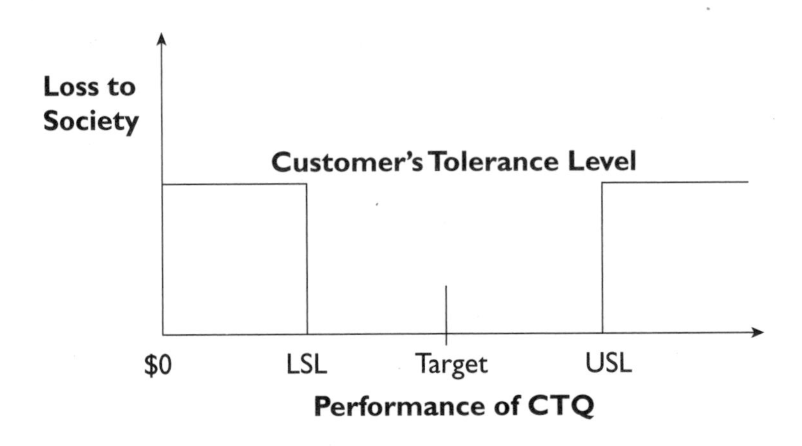

Figure 6-5 Traditional representation of a quality loss function.

lower specification limit (LSL) and the upper specification limit (USL) represent a customer's tolerance levels in the traditional approach.

Taguchi disagreed with it because there is no exception to "conformance to specifications" (1993). If the target characteristic deviates a short distance away from the set target, then in the long run it will cause societal loss because of customer dissatisfaction. The quality loss function includes the cost involved in filling the gap between desired and actual product quality.

Figure 6-5 shows zero Loss to Society between the tolerance limits, then a constant level outside the limits. The incorrect assumption for the traditional QLF is that financial loss is only incurred when you make scrap or when the out-of-spec product reaches the customer, probably incurring warranty costs. From the customer's viewpoint, there is probably very little difference between a product that is "just out of spec" and one that is "just in spec." It is unlikely that the customer could tell the two apart. By the traditional philosophy, one product is OK and the other is not. Also from the customer's viewpoint, the difference in quality between a product at the target and one that is just in spec could be small; however, there is the potential for a great difference in quality.

6.2.2 TAGUCHI'S QUALITY LOSS FUNCTION

Dr. Taguchi recommends that the goal should be to meet the target, not just to stay within the specification limits. The farther away from the target you are, the lower the quality and the greater the customer dissatisfaction. According to Dr. Taguchi's Quality Loss Function philosophy, as performance varies from the target, financial losses because of customer dissatisfaction increase—Loss to Society. He describes quality as, "the minimum loss imparted by the product to society from the time the product is shipped" (Taguchi, 1986).

The QLF is a mathematical model of monetary loss (due to suboptimal product performance) as a function of the deviation in the quality characteristic (response) from a target value. It includes the cost of lost opportunity as a result of customer dissatisfaction and loss of resources used to rectify the failure mode.

As shown in Figure 6-6, monetary losses can be approximated by a quadratic function of deviation of the response from target. Total loss may include costs related to assembly difficulty, degraded performance, customer dissatisfaction, warranty service costs, replacement or repair costs, and so on.

All products and processes experience variation. Based on the Quality Loss Function philosophy, the more variation in the distribution of a product's response (quality characteristic), the higher the losses. Systems with less response variability result in lower losses.

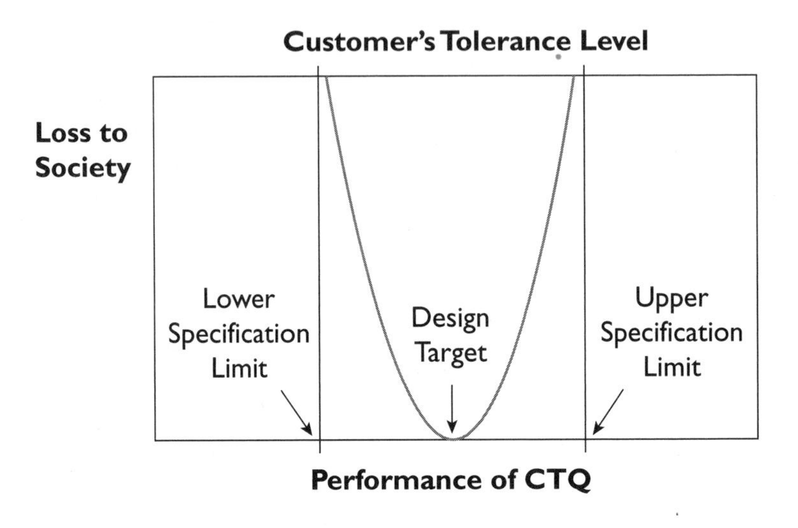

Figure 6-6 Taguchi's Quality Loss Function.

In sum, the QLF is a mathematical relationship between financial Loss to Society and deviation of the response from the target. The quality loss function takes into consideration all financial losses when a product deviates from its target, not just when it is turned in for warranty service.

6.3 NOISE FACTOR MANAGEMENT STRATEGIES

Fundamentally, robust design is concerned with minimizing the effect of uncertainty, or variation in parameters, on a design without eliminating the source of the uncertainty or variation. In other words, a robust design is less sensitive to variation in uncontrollable design parameters than traditional optimal designs. Robust design has found many successful applications in engineering and is continually being expanded to different design phases. An important goal for robust design is to minimize variation caused by uncontrollable noise by developing an effective noise factor management strategy.

6.3.1 IDENTIFYING NOISE FACTORS

As described in Section 6.1.3, sources of variability are called *noise.* Products and processes are subject to many sources of variation. Examples include customer usage, environmental considerations, deterioration, and manufacturing variability; these affect response variation and, hence, influence quality loss. The objective of the Parameter Design process is to achieve *robustness*—the quality of a product or process that provides its intended function in the presence of noise—with low variability and at minimal cost.

Lack of robustness is synonymous with excessive system variation, resulting in quality loss, which could be minimized if the system functioned on-target all of the time, despite the noise. For example, a robust starter would function consistently well regardless of the operating conditions and vehicle status. Customers expect cars to start every time, on the first turn of the key, regardless of the following:

- Ambient temperature
- Humidity
- Age of the car
- Altitude
- Hot or cold engine

Products and processes are subject to other sources of variation that can result in quality loss. In the context of robust design, noise is any source of variation causing quality loss. A robust product or process provides its intended function during its useful life (1) in the presence of noise,

(2) at minimal cost, and (3) with low variability. In other words, robustness is a state of functional performance that is insensitive to variations in raw materials, manufacturing processes, and operating environments at a cost that represents value.

6.3.2 STRATEGIES TO MANAGE NOISE

As discussed before, noise factors are the parameters that influence system, subsystem, and/or component variability and that can be difficult, expensive, or impossible to control. They disrupt energy transfer, causing it to be diverted from an intended function. Products or processes are subject to many sources of variation, including piece-to-piece, degradation, customer usage and/or duty cycle, environmental conditions (external), and system interfaces (internal). Many sources of noise are associated with products and processes, and there are several approaches for dealing with noise (e.g., ignore, compensate, reduce or eliminate, and/or minimize the impact). Some are more common than others.

Ignore

One of the traditional methods for dealing with noise is to ignore it. For example, you might use only perfect components when performing tests or you might conduct testing within a narrow range of customer usage conditions. Such practices do not take noise into account at all during the design phase. Some developers simply ignore noise factors and specifically remove them when testing during the development process.

Compensate

Another traditional method of handling noise is to compensate for it. This involves using feedback or feed-forward controls.

- *Feedback control*—the practice of sampling the output response and then changing an adjustment factor so that the response average moves back to on-target performance or feed-forward control

- *Feed-forward control*—the practice of evaluating the level of noise and then changing an adjustment factor so that the response average stays on-target

Reduce or Eliminate

A third traditional method for dealing with noise is to reduce or eliminate it. For instance, you might tighten tolerances or stipulate usage practices for a product. Preventive maintenance and common quality-control methods are ways to reduce or eliminate noise in a manufacturing environment. Example applications include (1) instructions for proper use of a product and (2) tightening tolerances.

Minimize the Impact

Developing products and processes in a perfect world is not practical and can result in products that do not function consistently in the real world. Eliminating, reducing, and compensating for noise are typically costly or time-consuming. It would be more practical to minimize the impact that any noise can have on the system being studied. Minimizing the effect of noise is the objective of Parameter Design (see also Chapter 7).

6.3.3 Parameter Design as a Noise Management Strategy

In Parameter Design, designed experimentation is used to identify (preferably at low cost) factor levels that will result in a system that is less sensitive to noise. The objective is to deal with variation by adjusting inexpensive and easy-to-change parameters. Using Parameter Design to manage noise has the following advantages over traditional countermeasures:

- It is less costly than compensating for, eliminating, or reducing noise.
- It is a more practical and proactive approach than ignoring noise.

EXAMPLE 6.1

The Cedar Tile Company had high piece-to-piece variability on its tile dimensions (see Figure 6-7). The interim solution was 100 percent inspection; however, the increase in staff time and the high level of scrap resulted in increased costs. (A similar example has been used by Dr. Taguchi to illustrate his method for robust design.)

To reduce inspection and scrap costs, a study was commissioned to find the cause of the variability. Results showed that there was a nonuniform temperature distribution that caused the tiles' variation. The tiles at the center of the pile inside the kiln were exposed to a lower temperature than those on the periphery.

Traditionally, engineers would put a lot of effort into attempts to control the cause or source of tile variation, including the following:

- Upgrade the technology by building a new kiln or adding more burners
- Feedback control of the temperature variation in the oven

- ➢ Differences in width and length
- ➢ Lack of parallelism
- ➢ Causes "cushion" or "trapezium"
- ➢ Labor-intense quality control (QC)
- ➢ QC performed under harsh industrial environment with noise, extreme temperatures, and high humidity

Figure 6-7 Cost of poor quality caused by tile-to-tile variation.

These approaches, however, are high-cost solutions. Controlling the cause of variation generally results in increased cost. Instead of using one of the traditional approaches, such as building a new kiln (at a cost of $1 million) or modifying the design of the kiln (at a cost dependent on the nature of the modification), Parameter Design was applied.

The Cedar Tile engineering team decided to conduct Parameter Design experiments using orthogonal arrays. Analysis concluded that increasing the lime content from 1 percent to 5 percent would greatly reduce the tile-size variation. The cost of increasing the lime content was minimal, especially when compared to the cost of a new kiln and the cost of inspection and scrap.

As shown in Figure 6-8, the tile company was able to solve the problem by minimizing the effect of the noise (varying temperatures) without controlling (modifying the kiln) or removing it (buying a new kiln); that is, the effect of nonuniform temperature distribution was minimized without controlling the temperature from within the kiln. Increasing the lime content made the tile robust to temperature variations within the kiln. The company was able to improve the robustness of the tile manufacturing process without increasing cost.

- Designed experimentation
- Solution: Add 4 percent more lime
- Size variation reduced
- Reduced amount of waste
- No additional costs

Figure 6-8 Increasing lime content reduces tile-size variation.

6.4 THREE PHASES: FROM THOUGHT TO THINGS

Dr. Taguchi recommends a three-phase approach to the development of robust products. For maximum efficiency and optimal system performance, the phases should be employed in the following order.

1. *Concept Design* involves selecting the appropriate level of technology required to provide a particular function.
2. *Parameter Design* involves determining nominal parameter settings that optimize the robustness of the selected technology at low cost. It is preferable to conduct Parameter Design before Tolerance Design because implementing the results of Parameter Design is less costly than those of Tolerance Design.
3. *Tolerance Design* involves selectively tightening tolerances and/or upgrading materials (based on a cost–benefit analysis) to achieve the required level of function.

Traditionally, engineers have focused more on Concept Design and on tightening tolerances, with less emphasis on Parameter Design; this can lead to suboptimal designs and/or excessive cost. Ideally, Tolerance Design should be conducted only when it is believed that Parameter Design cannot sufficiently improve robustness. When using Parameter Design, the most robust design can be determined by the following:

- Planning a designed experiment, including identifying the ideal function and the noise strategy
- Running the designed experiment with low-cost components or process settings
- Using two-step optimization
- Selecting parameter levels that reduce response variability due to the noise

- Adjusting the levels of other parameters that shift the mean of the response toward the target
- Selecting low-cost settings for those parameters with minimal effect on response variability or mean

As shown in Figure 6-9, the *response* is the measured value collected for each run of an experiment. Here experimental formulations that involve responses that can be categorized as either "static" or "dynamic" are considered. For maximal benefit, a response should be a metric that is continuous, objective, and a quality characteristic (as opposed to a measure of failure). Choosing a suitable response is a key issue in Parameter Design.

Static Parameter Design focuses on the improvement of a singular product or process, with a specific (singular) target. Such targets include zero, a particular nominal value, or even infinity (e.g., mean time between failures).

Run No.	A	B	C	D	E	F	G	Response
1	–	–	+	–	+	+	–	1.6
2	+	–	–	–	–	+	+	7.4
3	–	+	–	–	+	–	+	1.6
4	+	+	+	–	–	–	–	7.3
5	–	–	+	+	–	–	+	7.6
6	+	–	–	+	+	–	–	12.8
7	–	+	–	+	–	+	–	7.4
8	+	+	+	+	+	+	+	13.5

Figure 6-9 An experimental design matrix.

Dynamic Parameter Design involves studying a system across a range of its functionality.

Qualifying the experiment as static or dynamic depends on the nature of the response chosen for the system being studied. Although there are many similarities between static and dynamic Parameter Design, the analysis of dynamic formulations is slightly more complicated than the analysis of static formulations.

Traditionally, the focus of quality engineering has been on the improvement of a singular product or process, with a specific or singular target. An example would be a study to reduce the variability in the resistance of a 100-ohm resistor; static Parameter Design can be used for this.

A more beneficial approach involves gaining generic knowledge about a family of products or about a range of a product's applicability. With dynamic Parameter Design, the fundamental functional intent of the system is studied, and the system is evaluated across a range of functionality or applicability.

In sum, Concept Design selects an appropriate level of technology. Parameter Design optimizes the robustness of the technology selected. Tolerance Design selectively tightens tolerances and upgrades materials, if necessary, to achieve the required level of function. Response is the characteristic that determines whether a Parameter Design experiment is static or dynamic.

Based on P-Diagrams, an experimental design will help the product development team accomplish the following:

- Increase knowledge of a product or process
- Design a series of structured tests that are well planned, with changes on input variables
- Assess effects and interactions of input variations on predefined responses

6.5 SUMMARY

Development of the P-Diagram should begin alongside development of the Failure Mode and Effect Analysis (see Chapter 5). A Parameter Diagram enables the team to identify and review design specifications, as well as control and noise factors that affect the ideal function of their system.

Based on a P-Diagram, the robust design strategy, according to Dr. Taguchi, entails the maximization of the signal-to-noise ratio; that is, the minimization of the noise's impact with respect to the main factors' effect. This results in the creation of an understandable and well-defined system function in terms of measurable objectives. Figure 6-10 shows a flowchart for laying out a robust design strategy.

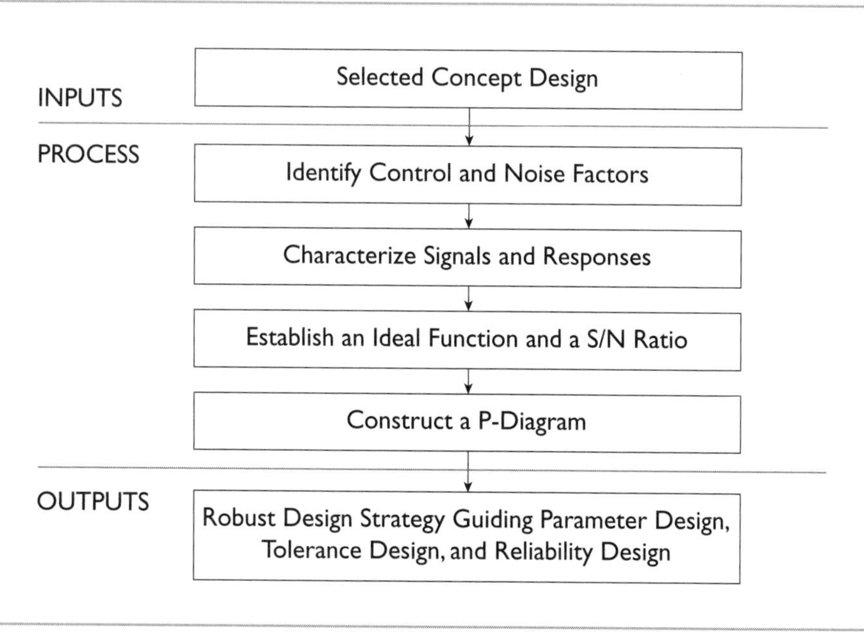

Figure 6-10 A flowchart for laying out a robust design strategy.

Experimental design is a systematic approach for identifying and solving engineering problems using organized, multifactor testing. It consists of a series of tests in which purposeful changes are made to the input variable of a process or system so that the reasons for change in the output response can be identified:

- Optimizing a design or process
- Selecting between different designs (processes) or design (process) concepts
- Understanding or identifying factors that affect function or performance
- Understanding how to make a design robust
- Validating robustness
- Validating computer models
- Developing engineering models
- Performing Parameter Design (Chapter 7)
- Performing Tolerance Design (Chapter 8)

BIBLIOGRAPHY

Box, George E. P., and Norman R. Draper. *Empirical Model-Building and Response Surfaces.* New York: John Wiley & Sons, 1987.

Box George E. P., William. G. Hunter, and J. Stuart Hunter. *Statistics for Experimenters—An Introduction to Design, Data Analysis, and Model Building.* New York: John Wiley & Sons, 1978.

Cuthbert, Daniel, and Fred S. Wood. *Fitting Equations to Data.* New York: John Wiley & Sons, 1980.

Lorenzen, Thomas, and Virgil Anderson. *Design of Experiments: A No-Name Approach.* New York: Marcel Dekker, 1993.

Montgomery, Douglas C. *Design and Analysis of Experiments, Fifth Edition.* New York: John Wiley & Sons, 2000.

Ray, Ranjit K. *Design of Experiments Using the Taguchi Approach.* New York: John Wiley & Sons, 2001.

Ross, Phillip J. *Taguchi Techniques for Quality Engineering.* New York: McGraw-Hill, 1995.

Taguchi, G. *Introduction to Quality Engineering.* Asian Productivity Organization, 1986; distributed in U.S. by American Supplier Institute, Dearborn, MI.

Taguchi, Genichi, and Yoshiko Yokoyama (Eds.). *Taguchi Methods: Design of Experiments.* Quality Engineering Series, Vol. 4. Dearborn, MI: American Supplier Institute, 1993.

Walters, Frederick H., Lloyd R. Parker, Jr., Stephen L. Morgan, and Stanley N. Deming. *Sequential Simplex Optimization.* Boca Raton, FL: CRC Press, 1991.

Winter, B. J. *Statistical Principles in Experimental Design.* New York: McGraw-Hill, 1971.

7 Parameter Design
Optimizing Control Factor Levels

Parameter Design is a principle that emphasizes choosing the proper levels for the controllable factors in a process for manufacturing products. When said to be optimal, the implication is that the design has achieved most of the target values set out by the quality measure before proceeding to a Tolerance Design. In an industrial setting, totally removing noise factors can be very expensive. Through a Parameter Design, engineers try to reduce the variation around the target by adjusting control factors' levels rather than by eliminating noise factors. By exploiting nonlinearity of products or systems, Parameter Design achieves robustness, measured by a signal-to-noise ratio (SNR, S/N), at a minimum cost. Orthogonal arrays are used to collect dependable information about control factors with a small number of experiments.

For the case study prepared for this chapter, a Six Sigma design team worked to develop a robust gyrocopter—a simple child's toy, made of paper, that presents some interesting and challenging design problems. The idea of designing for robustness, then tuning to target performance, is critical to robust design. This chapter also presents the following:

- How to verify critical inputs using a design of experiment (DOE)
- How to determine the optimal operating window for design parameters
- How to establish the relationship between Critical-to-Quality (CTQ) characteristics and control factors using regression analysis

Many of the topics related to planning and conducting a static Parameter Design experiment have already been discussed in Chapter 6. This chapter describes how to plan, conduct, and analyze a dynamic Parameter Design experiment.

A Parameter Design experiment is categorized as either static or dynamic based on the nature of the response. As such, although the step-by-step process for the development of each is nearly the same, several additional considerations are necessary when planning and analyzing dynamic Parameter Design experiments. Most engineering systems are dynamic, so whenever the project objective is to improve system robustness, dynamic Parameter Design should be applied—if feasible.

The *response* is the measured CTQ value collected for each run of an experiment. Experimental formulations that involve responses that can be categorized as either static or dynamic are considered in this chapter. Chapter 6 noted that, for maximal benefit, a response should be a metric—that is, continuous, objective, and a quality characteristic—as opposed to a measure of failure. Choosing a suitable response is a key issue in Parameter Design. Determining what to measure—the response—in your experiment is of fundamental importance. For maximal benefit of the application of Parameter Design, you should measure a dynamic response whenever possible.

A *dynamic response* is a characteristic that increases (along a continuous scale) in proportion to a related input to the system—called the signal. In dynamic Parameter Design formulations, the response is studied at vari-

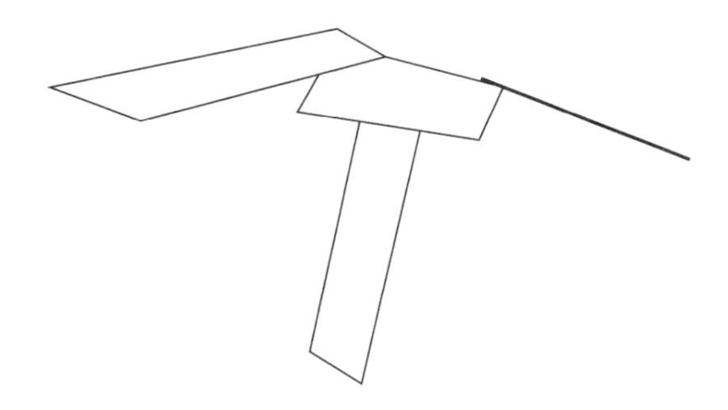

Figure 7-1 A gyrocopter's flight time (dynamic response) increases with drop height (signal).

ous signal levels. For the paper gyrocopter shown in Figure 7-1, the dynamic response (flight time) increases with the signal (drop height).

Because the system is studied at various signal levels, information is available to optimize the design throughout its intended range of operation; this creates flexible and reproducible technology. Of course, to achieve robustness, the response must also be studied across noise conditions, as in static Parameter Design. The following seven-step process is recommended to organize the development of a dynamic Parameter Design experiment:

1. Identify project and team
2. Formulate ideal function
3. Formulate dynamic Parameter Design
4. Assign control factors to an inner array
5. Assign noise factors and signal to an outer array
6. Conduct the experiment and collect data
7. Analyze the data and select an optimal design

As a result of doing systematic experimentation, using sound statistical principles, the quality of processes can be improved and become more robust to variations in component and processing factors' levels, as described in the sections that follow.

The key differences between static and dynamic Parameter Designs will become apparent in Steps 2, 5, and 7. Step 2 considers the physics-based relationship between the signal and the response, called the *ideal function.* Step 5 shows how to accommodate a dynamic response in the experimental plan, and Step 7 details how to analyze dynamic response data.

7.1 IDENTIFY PROJECT AND TEAM

The characteristics of a successful project are a clear objective, including the desired outcome; a cross-functional team that includes suppliers; thorough planning; and management support. Sponsorship and support are critical for success. Management's role is to provide necessary resources, empower the team, and remove obstacles to progress.

The first step in dynamic Parameter Design is to identify the project and the team. As with any project, effective planning and selection of the right team members can make the difference between success and failure. Project selection should be based on the potential for increasing customer satisfaction, increasing reliability, incorporating new technologies, reducing cost and warranty service, and achieving best-in-class quality.

GYROCOPTER CASE STUDY EXAMPLE

The gyrocopter case study is used as a practice exercise throughout this chapter. You are a Six Sigma Black Belt working at Marion High-Tech Fun for Kids; its Web site includes cut-out patterns of various flying objects to be assembled by five- to ten-year-old children.

Although customer surveys suggest that some older children have been known to build a few of these contraptions, reportedly they only did so, of course, when providing consultation to five- to ten-year-olds! As shown in Figure 7-2, the *children* obtain a copy of the pattern from their printers, cut it out, fold it to produce the flying object, and then launch the gyrocopter. Customer surveys have revealed disappointment with the copters made from the pattern shown in the figure.

Conclusions drawn from surveys suggest that customers' highest priority is the ability of the gyrocopter to provide consistent flight. They also like long flight times, but the children are especially disappointed if the copters vary in their ability to consistently deliver good flight times, both from drop-to-drop and from build-to-build. Their expectation is that if they carefully build the design exactly as given and launch it in the recommended manner, it will work well—as well as a previously built one and, more important, as well as the one built by the kid next door. The team might include:

- A seasoned gyrocopter engineer
- A new gyrocopter engineer
- A flight-testing technician
- Someone from the customer survey department to provide insight into how customers use (cut, build, and launch) the copters and their relative expectations

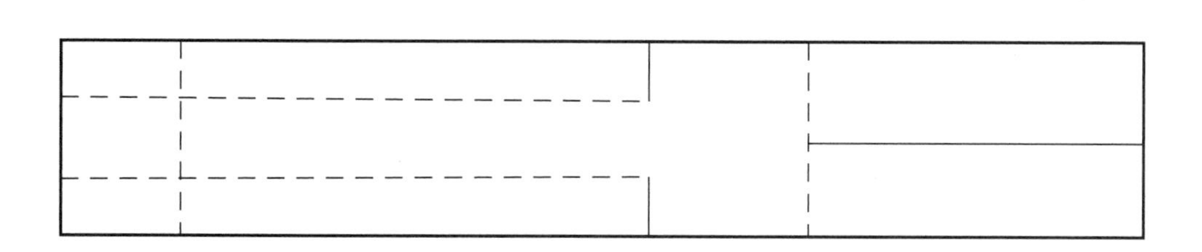

Figure 7-2 A gyrocopter pattern.

The objective is to design a gyrocopter that consistently provides reasonably long flight times. Since consistency is a big issue, it is important to produce reasonably long flight times rather than simply maximizing the mean of the distribution of flight times. Presumably such a distribution includes many long flight times; however, without a focus on improving consistency (reducing the variation) such a distribution has the potential to also include many short flight times, thus failing to meet customers' requirements.

7.2 FORMULATE IDEAL FUNCTION

The second step in dynamic Parameter Design is to formulate the engineered system's *ideal function*, which is a mathematical description of its fundamental intent. It is the physics-based, functional relationship between the signal and the response. The ideal function represents perfect (ideal) functionality of the system, and so might be considered the goal of the experiment. In the simple example, the signal is the height from which the gyrocopter is dropped and the response is its flight time.

Customer requirements are the starting point for determining what to measure in an experiment. But customer performance metrics are sometimes vague, usually subjective, and typically expressed in nontechnical terms. So, to produce quality products, the engineer must translate customer performance metrics into measurable, objective, engineering metrics.

The Voice-of-Customers (VOCs) conveys to the engineer what customers want and how they perceive what they actually get. The VOC is composed of customer intent and perceived results. *Customer intent* is what the customer wants or expects; the *perceived result* is customers' perception of what they get. Together, these represent the customers' world. When the perceived result doesn't match the intent, customers are disappointed.

The means by which customers use the system to fulfill their intent serves only as a starting point for examining the system's intended function. The engineer must then study the physics of the system to identify engineering metrics (the signal and the response) that relate to the functional intent. The metrics are denoted by the variables M and Y, respectively. In general, the signal (M) is the level of input received by the system, and the response (Y) is a measure of the functional outcome.

More specifically, the signal is an engineering variable that quantifies the magnitude of input to the system—in the example, it is drop height in feet. The signal initiates energy transformations within the system, producing the response. The response is an engineering metric that quantifies the functional performance of the system, such as flight time in seconds.

As shown in Figure 7-3, the linear relationship between signal and response is called the system's ideal function. The signal is the independent variable,

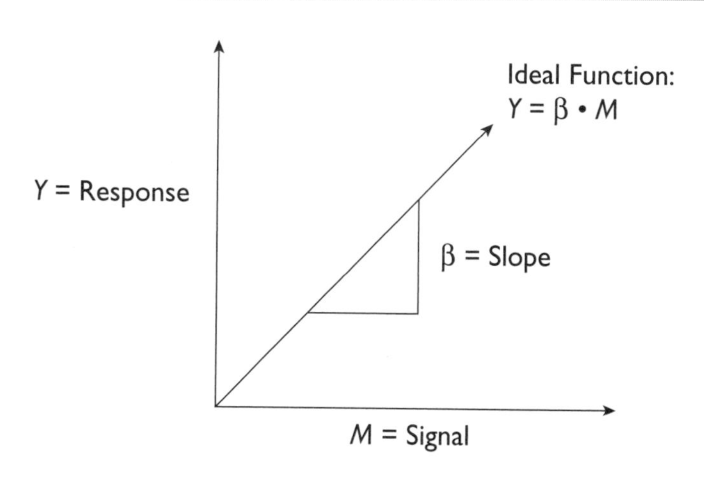

Figure 7-3 The ideal function.

plotted along the horizontal axis. The response is the dependent variable, plotted along the vertical axis. The slope of the ideal function line is denoted by β—often referred to as the sensitivity—and is used in Dr. Taguchi's dynamic signal-to-noise ratio. Since the signal and the response should be related to energy transfer through the system, the ideal function should pass through the origin (no energy in, no energy out).

Sometimes, a transformation or a substitution of the variables is necessary to derive the linear ideal function. For nonlinear relationships (e.g., quadratic or exponential relationships), the M or Y axis can be appropriately transformed to produce a linear relationship.

GYROCOPTER EXAMPLE CONTINUED

Although this is probably a bit of a challenge without the support of your team, and in a very short time frame, it is time to take a shot at developing the ideal function for your gyrocopter dynamic Parameter Design project.

Before embarking on this task, take the time to consider the function of the paper gyrocopter. Also decide whether you think that it makes sense to focus on a subsystem or on the entire copter. Assuming that your team will tackle the entire system, you need to determine each of the aspects of the engineered system shown in Figure 7-4.

Your team then dedicates many hours to the development of the ideal relationship between the signal (drop height) and the response (flight time). After much discussion, the team agrees on the following:

$$\text{Flight time} = k \cdot \text{Drop height}$$

Although this relationship may seem trivial at first glance, its derivation is actually rather sophisticated. It is based on the assumption that the gyrocopter would function "ideally" if it began rotating and hit terminal velocity immedi-

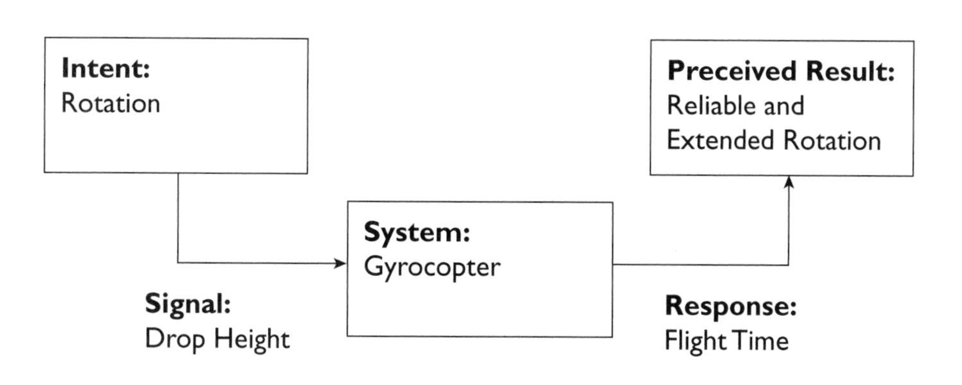

Figure 7-4 Forming the ideal function of a gyrocopter.

ately. If this happened, the copter would descend at a constant velocity for its entire flight, suggesting a linear relationship between flight time and drop height, regardless of drop height.

Keep in mind that if the gyrocopter had been an object in free fall, the relationship would have been nonlinear, but the ideal function is based on the ideal theoretical functionality; it represents the *goal* for the system.

7.3 FORMULATE DYNAMIC PARAMETER DESIGN

The *engineered system* for a product or process shows the relationship among the particular system or subsystem, the response, and the associated control and noise factors. The process of creating a system is the same for static and dynamic Parameter Design.

Most control factors are obvious to the engineer because they relate directly to system design. On the other hand, it is easy to overlook some noise factors because they are frequently external to system design. As discussed in Chapter 1, the following are five potential sources of noise:

- Internal environment, due to neighboring subsystems
- Part-to-part (piece-to-piece) manufacturing variation
- Customer usage and duty cycle
- External environment
- Age, or deterioration

It is important to identify all of the factors with labels, as shown in Figure 7-5. Examining the five potential sources of noise based on a diagram can help engineers develop a thorough list of noise factors. Table 7-1 shows a brainstormed list of control and noise factors for the gyrocopter system. Your list might include some of these items.

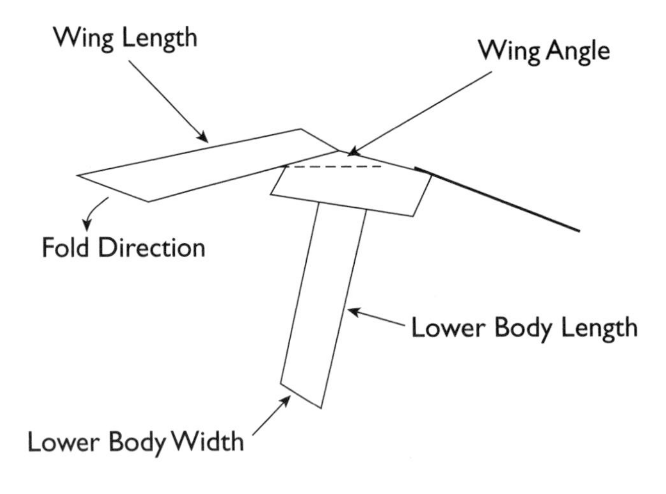

Figure 7-5 A labeled diagram for a gyrocopter.

Table 7-1 Brainstormed List of Control and Noise Factors

ambient temperature
breeze
center of gravity
launch technique
lower body area
lower body fold direction
lower body length
lower body width
paper weight and type
tail fold
taped body crease
upper body area
upper body length
upper body shape
upper body width
wing angle with respect to the indicated fold line
wing angle fold line with respect to the indicated fold line
wing crease stiffness
wing flaps
wing fold direction
wing gussets
wing length
wing width

Brainstorming is a problem-solving approach or technique whereby working members in a group conduct a deductive methodology for identifying possible factors to improve performance in any product or process pursued by the members and facilitator. A benefit of brainstorming is the power of the group to build ideas based on each other's ideas. The following are elaborations on several items shown in Table 7-1:

- Wing width is a control factor.
- Wing crease stiffness is a noise factor.
- Fold direction and paper type can be noise factors or control factors, depending on the perspective of the team. If the team decides to select a certain fold direction and paper type through experimentation, then they are control factors; otherwise, they are noise factors.
- Tail fold is a control factor. This design characteristic, like any of them, could easily be ignored by customers; but the assumption is that they will build a copter as indicated on the pattern, so it is probably a control factor.

7.4 Assign Control Factors to an Inner Array

Once the ideal function and associated noise and control factors have been identified, the team can begin to develop the experimental plan. In a Parameter Design experiment, the plan consists of two arrays—the inner array and the outer array. Control factors are assigned to the inner array, and the noise and signal factors are assigned to the outer array.

Based on their relative impact on the system and on available resources, the team must determine which control factors to include in the experiment. They must then identify the level for each factor and assign the factors and levels to the inner array. This process is the same for both static and dynamic Parameter Design.

GYROCOPTER EXAMPLE CONTINUED

Due to timing constraints, the team decides to use an L_8 inner array, as shown in Figure 7-6. Because the team members think that some control factors may interact, they elect to test only four of them, each at two levels. Further, previous studies have suggested that, to remain upright, lower body length (LBL) must exceed wing length (WL) in most designs and that these two factors interact. To account for this, the team decides to set the LBL levels relative to wing length, as shown in Figure 7-7.

The L_8 orthogonal array used in the figure has eight experimental runs (rows) and can accommodate up to seven factors (columns). It is not necessary to use all of the columns. In this experiment, because just four factors are being investigated, only four columns are used.

As a result of this, the experiments can be conducted with a smaller number of runs; that is, instead of evaluating the four control factors at 2^4 (= 16 combinations), the procedure uses an L_8 orthogonal array. Using orthogonal arrays

Run No.	Col. 1	Col. 2	Col. 3	Col. 4	Col. 5	Col. 6	Col. 7
1	1	1	1	1	1	1	1
2	1	1	1	2	2	2	2
3	1	2	2	1	1	2	2
4	1	2	2	2	2	1	1
5	2	1	2	1	2	1	2
6	2	1	2	2	1	2	1
7	2	2	1	1	2	2	1
8	2	2	1	2	1	1	2

Figure 7-6 An L_8 inner orthogonal array.

Control Factors	Level 1	Level 2
Wing length (WL)	4	10
Lower body length (LBL)	1.25 x WL	2 x WL
Upper body shape (UBS)	Angle	Square
Tail fold (TF)	None	3

Run No.	WL	LBL	Col. 3	UBS	Col. 5	Col. 6	TF
1	1	1	1	1	1	1	1
2	1	1	1	2	2	2	2
3	1	2	2	1	1	2	2
4	1	2	2	2	2	1	1
5	2	1	2	1	2	1	2
6	2	1	2	2	1	2	1
7	2	2	1	1	2	2	1
8	2	2	1	2	1	1	2

Figure 7-7 Control factors selected by the gyrocopter robust design team.

reduces the number of function evaluations needed, which is very important when experiments are employed to predict performance characteristics. Here a 50 percent saving in computer analysis time is possible.

As shown in the Figure 7-8, an L_8 orthogonal array is often used to estimate the effect of four factors. If no three-factor interaction is expected (generally a good assumption), the fourth factor, *d*, may be represented by the column otherwise used for the *abc* interaction. Note that an estimate of the *bc* interaction is now confounded with the *ad* interaction; that is,

Taguchi's L_8 (2^7) Array

a	b	–ab	c	–ac	–bc	abc
1	1	1	1	1	1	1
1	1	1	2	2	2	2
1	2	2	1	1	2	2
1	2	2	2	2	1	1
2	1	2	1	2	1	2
2	1	2	2	1	2	1
2	2	1	1	2	2	1
2	2	1	2	1	1	2

Note: Used to investigate the effects of up to 7 factors in 8 runs.

Figure 7-8 Confounding when using L_8 for analyzing four factors.

they cannot be independently estimated. In Figure 7-7, the four control factors are allocated to columns 1 (WL), 2 (LBL), 4 (UBS), and 7 (TF). This Resolution III design of experiment will keep all of the main effects free from confounding with any two-way interactions. Factor or interaction effects are said to be *confounded* when the effect of one factor is combined with that of another. In other words, the effects of multiple factors on a response cannot be separated. Confounding occurs to some degree in all situations and least frequently when the data is obtained from a carefully planned and executed experiment having a predefined objective.

Resolution is a measure of the degree of confounding among effects; Roman numerals are used to denote it. The resolution of a DOE defines the amount of information that can be provided by the experiment's design. In general, the resolution of a design is one more than the smallest-order interaction that some main effect is confounded with. If some main effects are confounded with some two-level interactions, the resolution is III. Alias is the lost interactions in a design of experiment. The term *alias* indicates that two or more things have been changed at the same time in the same way. Alias is a critical feature of Taguchi designs or standard fractional factorials.

7.5 ASSIGN NOISE FACTORS AND SIGNAL TO AN OUTER ARRAY

The objective of dynamic Parameter Design is to identify a design that is insensitive to noise across the intended range of functions. To achieve this objective, the noise and signal must be included in the experimental plan. The outer array specifies the combination of noise and signal levels at which each control factor combination is tested.

Noise causes energy to be diverted away from producing the intended function. In the example, wing crease stiffness has been identified as a noise factor (NF) because the less rigid the crease is, the more likely the gyrocopter is to fall quickly. It is included in the experimental plan to create a testing environment that mimics potential usage conditions. In the static case, there is a single target value for the response. (It may be that the data is collected at a single signal level, M.) The asterisks (see Figure 7-9) indicate the ideal response (the target value). The optimal design exhibits minimal variability around the target.

At each of the signal levels in dynamic experimentation, data is collected under the influence of noise extremes. The goal (target) is the ideal function. The asterisks indicate the ideal response at each signal level. The optimal design exhibits minimal variability around the ideal function line. In the example in Figure 7-9, data has been collected for a given design at three signal levels (drop heights)—M_1, M_2, and M_3. Two responses (flight times) were collected at each signal level under the influence of N_2—the high-response condition; plus, two responses were collected at each sig-

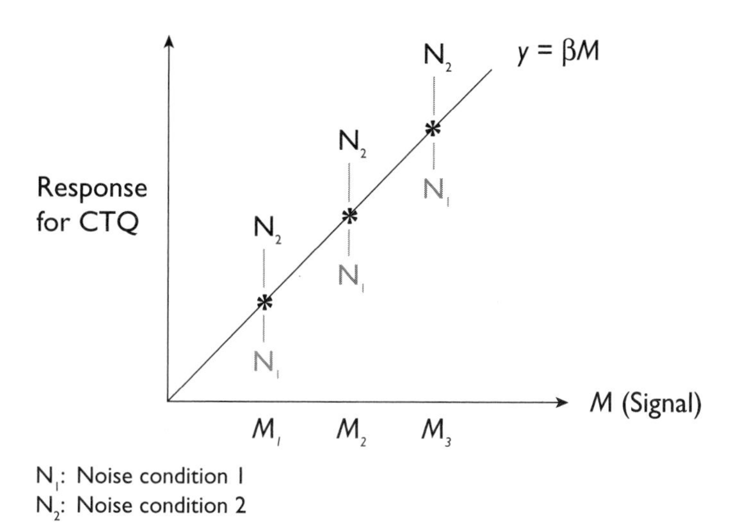

Figure 7-9 Impact of noise on a dynamic response.

nal level under the influence of N_1—the low-response condition. Noise can also be responsible for nonlinear system function. In this context, nonlinearity can be thought of in the same context as lack of repeatability, that is, as deviation away from the ideal straight line.

An efficient strategy for testing noise should be developed, especially if there are several strong noise factors. To accomplish this, first identify the most influential noise factors (fold direction, crease stiffness) and the noise levels that produce an extreme response. Then determine a strategy for testing at extreme noise conditions. This process is the same for both static and dynamic Parameter Design.

Once the noise strategy and the signal levels have been determined, they can be allocated to the outer array—the orthogonal array that specifies noise and signal treatment combinations. Each combination of control factor levels specified in the inner array is tested at each of the treatment combinations specified in the outer array.

To analyze the data from a dynamic Parameter Design experiment using the methods introduced by Dr. Taguchi, it is necessary to assign the noise and signal to an outer array. Signal levels (M_1, M_2, . . ., M_n) are the levels of input used during the experiment to generate the corresponding responses. Choose the levels based on both engineering judgment and selection guidelines. Testing across a broad signal range will push the technology to produce a response even at extreme signal levels.

Noise is included in the experimental plan to induce response variability. The signal is included in the experimental plan so that the design can be optimized for robustness across its intended and anticipated range of operation. At each signal level in the outer array, every noise condition is tested.

GYROCOPTER EXAMPLE CONTINUED

Suppose that the following noise factors and levels were proposed for the gyrocopter case study.

Wing crease stiffness:

- Rigid crease
- Stressed crease

Paper clip on tail:

- Heavy clip
- Light clip

Based on the proposed noise factors, the following compounded noise strategy is recommended:

- N_1 = short time (high velocity): stressed crease, heavy clip, heavy paper
- N_2 = long time (low velocity): rigid crease, light clip, light paper

Traditionally, this design has included a paper clip on the tail to help the gyrocopter establish and maintain its vertical position, and to help initiate rotation. As you might anticipate, the weight of the clips used by kids varies greatly. In fact, many times they use no clip at all. Moreover, customer usage surveys on this and other designs suggest that when wing creases become overstressed, the kids typically create fresh paper copters. So, the team decides that wing crease stiffness should not be included in their noise strategy.

Additionally, surveys have revealed that the vast majority of children make their copters out of printer paper. Noise tests suggest that the range of standard printer paper weights causes little separation in drop times. Therefore, the team eliminates the variation in paper weight as a noise factor and concludes that the only strong source of noise is the paper clips' weight. They decide to test this noise factor at two levels:

- N_1 = no paper clip
- N_2 = standard metallic paper clip

Based on the average height of a five-year-old and allowing for the possibility that the gyrocopters might be dropped from balconies or stairwells (or bedroom windows or trees, etc.), the team settles on the following signal levels:

- M_1 = 3 feet
- M_2 = 8 feet
- M_3 = 13 feet

Use N_1 and N_2 and M_1, M_2, and M_3 to identify the noise and signal level combinations that make up the outer array. M_1N_1, M_1N_2, M_2N_1, M_2N_2, M_3N_1, and M_3N_2 are the six combinations that should be tested. The inner and the outer arrays are shown together in Figure 7-10.

Run No.	WL	LBL	Col. 3	UBS	Col. 5	Col. 6	TF	M_1 = 3 feet		M_2 = 8 feet		M_3 = 13 feet	
								N_1 no clip	N_2 clip	N_1 no clip	N_2 clip	N_1 no clip	N_2 clip
1	1	1	1	1	1	1	1						
2	1	1	1	2	2	2	2						
3	1	2	2	1	1	2	2						
4	1	2	2	2	2	1	1						
5	2	1	2	1	2	1	2						
6	2	1	2	2	1	2	1						
7	2	2	1	1	2	2	1						
8	2	2	1	2	1	1	2						

Figure 7-10 The inner and the outer arrays.

7.6 CONDUCT THE EXPERIMENT AND COLLECT DATA

Even though the experimental plan is now complete, some preliminary considerations should be considered before running the experiment. After addressing these issues, testing can commence. Collect the data in the most convenient order and be sure to complete the experiment.

If the team has not already done so, members must now investigate available testing options. Once the means of testing is established, concise definitions of the factors and the levels need to be established; preliminary tests should be conducted; and, finally, experimental logistics can be determined. Before conducting the entire experiment as designed, it is recommended that you make a preliminary run for data (across all signal and noise combinations) on a baseline design, which might be the current design.

Since (at most) a couple of designs will be evaluated during preliminary testing, repetition is recommended. Construct scatterplots of the initial run data using different symbols for each noise level to determine that the noise strategy separates the responses across all signal levels and to evaluate the experimental procedures.

Frequently, it is more difficult or expensive to change the signal, the noise conditions, or the designs from run to run. Collect data in the order that requires the fewest number of changes for the CTQ that is the most difficult to reset. As shown in Figure 7-11, if it is difficult and/or expensive to change the noise conditions, you can do the following:

- First, collect all of the data at the N_1 condition across all signal levels and for all designs (runs).
- Next, change to the N_2 condition and collect the remaining data.

Randomization, of course, is also an option, although an effective noise strategy should make this unnecessary. All designs specified in the inner

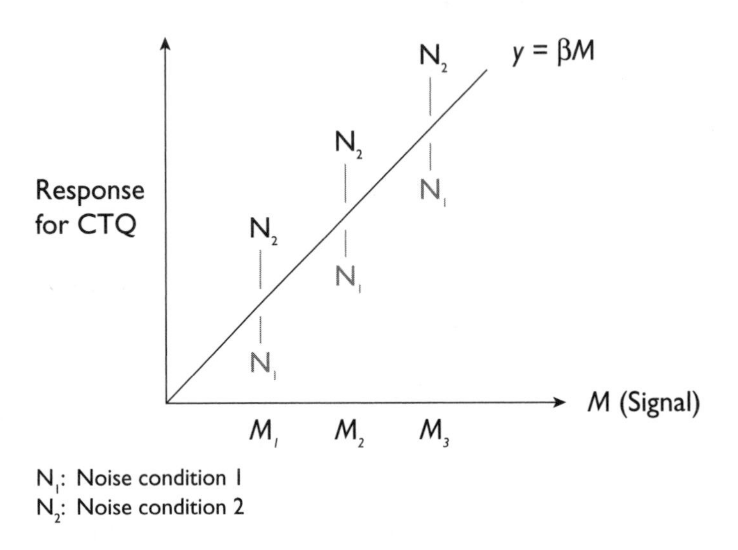

Figure 7-11 Scatterplot of preliminary data to check noise separation.

array must be tested in order to maintain balance. Without balance, the ensuing analysis approach will be invalid; all effect estimates will be biased. If unforeseen circumstances require a reduction in the size of the experiment, postpone data collection for either a noise condition or a signal level, bearing in mind that (1) ideally three signal levels should be included, but two will suffice if three is not feasible; and (2) to become robust to noise, noise extremes must be tested.

GYROCOPTER EXAMPLE CONTINUED

Based on the experimental plan, the Six Sigma Black Belt is running the experiment (see Figure 7-12). The flight time is measured in seconds. He is now

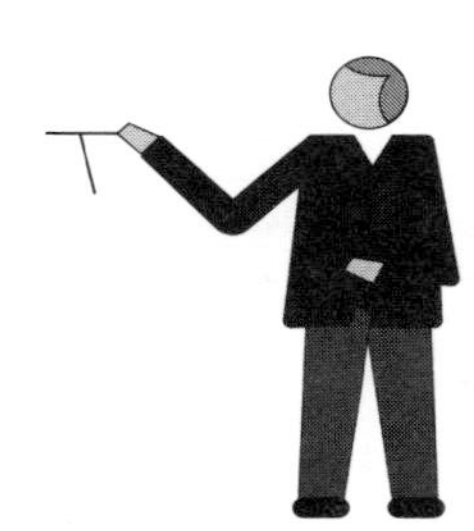

								M_1 = 3 feet		M_2 = 8 feet		M_3 = 13 feet	
Run No.	**WL**	**LBL**	**Col. 3**	**UBS**	**Col. 5**	**Col. 6**	**TF**	**N_1 no clip**	**N_2 clip**	**N_1 no clip**	**N_2 clip**	**N_1 no clip**	**N_2 clip**
1	1	1	1	1	1	1	1	0.7	0.5	2.0	1.4	3.0	2.1
2	1	1	1	2	2	2	2	0.7	0.4	2.0	1.0	2.9	1.9
3	1	2	2	1	1	2	2	0.6	0.5	1.7	1.2	2.7	2.0
4	1	2	2	2	2	1	1	0.7	0.4	1.7	1.1	2.9	2.0
5	2	1	2	1	2	1	2	1.2	0.9	3.4	2.3	5.3	3.7
6	2	1	2	2	1	2	1	1.2	0.8	3.0	1.8	5.1	3.3
7	2	2	1	1	2	2	1	1.1	0.8	2.8	2.0	4.5	3.3
8	2	2	1	2	1	1	2	1.0	0.7	2.8	1.8	4.5	**3.1**

Figure 7-12 Running the gyrocopter experiment.

obtaining the data for the Run 8/M_3/N_2 cell, which corresponds to the following conditions:

- WL – Level 2: 10
- UBS – Level 2: Square
- TF – Level 2: 3
- LBL – Level 2: 2 x WL
- N_2 – Clip
- M_3 – 13 feet

7.7 ANALYZE THE DATA AND SELECT AN OPTIMAL DESIGN

The overall analysis strategy is to use a two-step optimization process. The first step is to reduce variability and the second is to adjust sensitivity to the target. The dynamic signal-to-noise ratio is used to identify factor levels that make the system less variable; the slope of the best-fit line is used to adjust sensitivity. (Calculate the slope of the best-fit line using the formula shown in Figure 7-13.)

Analysis begins with the calculation of the slope, β, and the variance, σ, for each run of the experiment. With the ideal function, $y = \beta M$, as the goal, the formulas for calculating β and σ for each run are based on fitting a line through the origin to the data for that run. A common method of determining a best-fit line is called the *method of least squares*, which determines the line that minimizes the sum of the squared residuals. For each run, β is the slope of the best-fit line using this method, and σ^2 is a measure of variation around the $y = \beta M$ line for that run (Figure 7-15).

With mathematics, it can be shown that the line (through the origin) with the slope, β, as just given, will have a minimum sum of squared

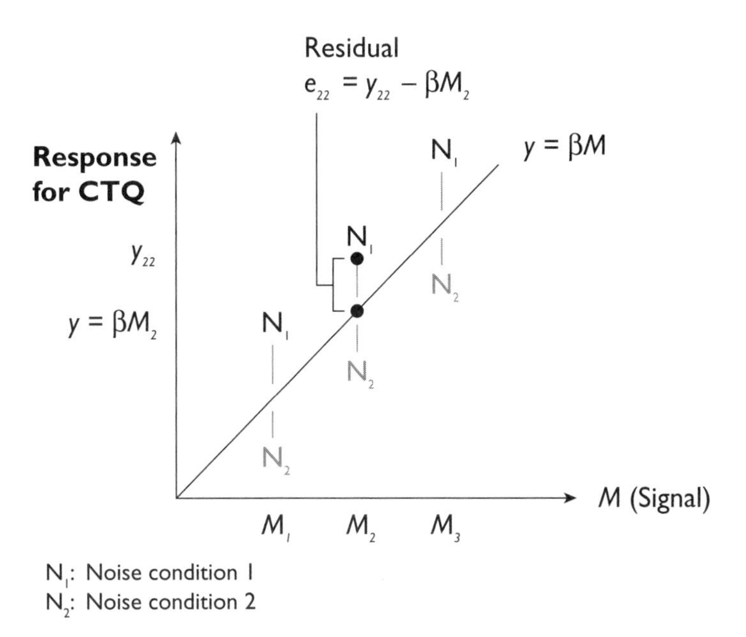

Figure 7-13 A best-fit line for each run.

residuals. An equivalent formula for calculating the value of β is given by this equation:

$$\beta = \frac{\sum_{i=1}^{m}\sum_{j=1}^{n} M_i y_{ij}}{n\sum_{i=1}^{m} M_i^2}$$

where m is the number of signal levels and n is the number of noise levels. If the experiment includes repeated data at each noise–signal level combination, the formula must be modified slightly.

The variation around the line with the least squares method is called the mean squared error (MSE) or, sometimes, the mean squared deviation (MSD). The formula for calculating σ^2 is as follows:

$$\sigma^2 = \frac{\sum_{i=1}^{m}\sum_{j=1}^{n}(y_{ij} - \beta M_i)^2}{mn-1}$$

where, again, m is the number of signal levels and n is the number of noise levels.

To calculate σ^2 for a given run, note that β must be calculated first. Then, given b, M is the point on the best-fit line (for that run) at signal level M, and $y_{ij} - \beta M_i$ is the deviation between the actual data point and the point on the best-fit line. This deviation is called the residual associated with y_{ij}. If the experiment includes repeated data at each noise–signal level combination, the formula must be modified slightly.

GYROCOPTER EXAMPLE CONCLUSION

Using the indicated formula, the β is calculated in Figure 7-14 and the σ^2 is calculated in Figure 7-15. Once β and σ have been determined, the dynamic signal-to-noise (S/N) value can be calculated for each run. The S/N can be thought of as a measure of relative variability. The dynamic signal-to-noise value is calculated as follows:

$$\text{S/N} = 10 \bullet \log\left(\frac{\beta^2}{\sigma^2}\right).$$

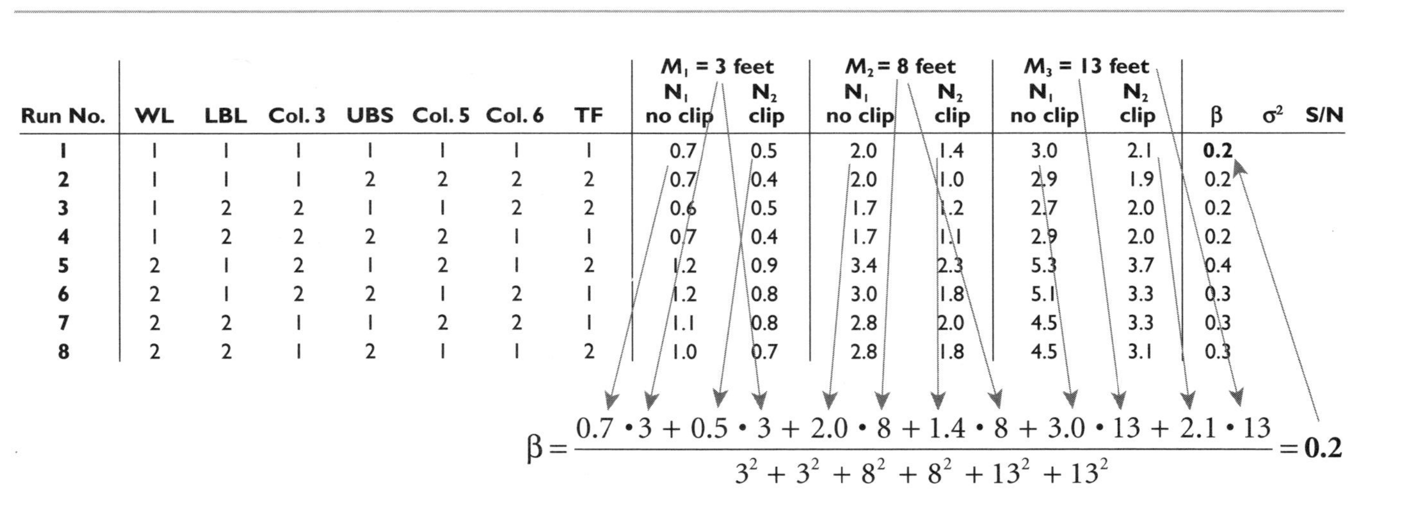

Run No.	WL	LBL	Col. 3	UBS	Col. 5	Col. 6	TF	M_1 = 3 feet N_1 no clip	M_1 = 3 feet N_2 clip	M_2 = 8 feet N_1 no clip	M_2 = 8 feet N_2 clip	M_3 = 13 feet N_1 no clip	M_3 = 13 feet N_2 clip	β	σ^2	S/N
1	1	1	1	1	1	1	1	0.7	0.5	2.0	1.4	3.0	2.1	**0.2**		
2	1	1	1	2	2	2	2	0.7	0.4	2.0	1.0	2.9	1.9	0.2		
3	1	2	2	1	1	2	2	0.6	0.5	1.7	1.2	2.7	2.0	0.2		
4	1	2	2	2	2	1	1	0.7	0.4	1.7	1.1	2.9	2.0	0.2		
5	2	1	2	1	2	1	2	1.2	0.9	3.4	2.3	5.3	3.7	0.4		
6	2	1	2	2	1	2	1	1.2	0.8	3.0	1.8	5.1	3.3	0.3		
7	2	2	1	1	2	2	1	1.1	0.8	2.8	2.0	4.5	3.3	0.3		
8	2	2	1	2	1	1	2	1.0	0.7	2.8	1.8	4.5	3.1	0.3		

$$\beta = \frac{0.7 \cdot 3 + 0.5 \cdot 3 + 2.0 \cdot 8 + 1.4 \cdot 8 + 3.0 \cdot 13 + 2.1 \cdot 13}{3^2 + 3^2 + 8^2 + 8^2 + 13^2 + 13^2} = \mathbf{0.2}$$

Figure 7-14 The β calculation done by the gyrocopter design team.

Run No.	WL	LBL	Col. 3	UBS	Col. 5	Col. 6	TF	M_1 = 3 feet N_1 no clip	N_2 clip	M_2 = 8 feet N_1 no clip	N_2 clip	M_3 = 13 feet N_1 no clip	N_2 clip	β	σ^2	S/N
1	1	1	1	1	1	1	1	0.7	0.5	2.0	1.4	3.0	2.1	0.2	**0.13**	
2	1	1	1	2	2	2	2	0.7	0.4	2.0	1.0	2.9	1.9	0.2	0.21	
3	1	2	2	1	1	2	2	0.6	0.5	1.7	1.2	2.7	2.0	0.2	0.08	
4	1	2	2	2	2	1	1	0.7	0.4	1.7	1.1	2.9	2.0	0.2	0.13	
5	2	1	2	1	2	1	2	1.2	0.9	3.4	2.3	5.3	3.7	0.4	0.39	
6	2	1	2	2	1	2	1	1.2	0.8	3.0	1.8	5.1	3.3	0.3	0.49	
7	2	2	1	1	2	2	1	1.1	0.8	2.8	2.0	4.5	3.3	0.3	0.22	
8	2	2	1	2	1	1	2	1.0	0.7	2.8	1.8	4.5	3.1	0.3	0.31	

$$\sigma^2 = \frac{(0.7-0.2-3)^2+(0.5-0.2\cdot 3)^2+(2.0-0.2\cdot 8)^2+(1.4-0.2\cdot 8)^2+(3.0-0.2\cdot 13)^2+(2.1-0.2\cdot 13)^2}{3\cdot 2-1} = \mathbf{0.13}$$

Figure 7-15 The σ^2 calculation done by the gyrocopter design team.

Increasing β or decreasing σ results in higher S/N. So, just as in static Parameter Design, to reduce (relative) variability, maximize S/N. The units for β depend on the units for M and y. The units for σ are the same as the response units. Units for S/N are always in decibels, which are calculated for the gyrocopter as follows: For each run, calculate the dynamic signal to noise ratio. As with all S/N formulations, for maximal robustness, the ratio needs to be maximized regardless of the nature of the response.

A variable that represents the performance of a product or a process is called a *performance characteristic.* A performance characteristic deviates from its target value because of noise such as manufacturing variation, environmental conditions, or deterioration. For most engineered products, the target values of a performance characteristic could vary from time to time according to the requirements of customers.

Such a characteristic is called a *dynamic characteristic,* and the input variable used to attain the required target values is called a *signal parameter.* The goal of dynamic Parameter Design for a dynamic characteristic problem is to achieve the smallest variation of the characteristic around its target values over the range of the signal parameter under various noise conditions.

The objective of Parameter Design is to improve product or process performance by determining the levels of its design parameters such that the performance characteristic is robust against various causes of variation. For this purpose, Dr. Taguchi developed an experimental method in which orthogonal arrays are used as experimental designs and a SNR is employed for analyzing the data. According to Dr. Taguchi, failures are caused either by an excess of or a deficiency in energy. Therefore, efficient products produce fewer failures than inefficient products. The process for conducting a dynamic Parameter Design is shown in Figure 7-16.

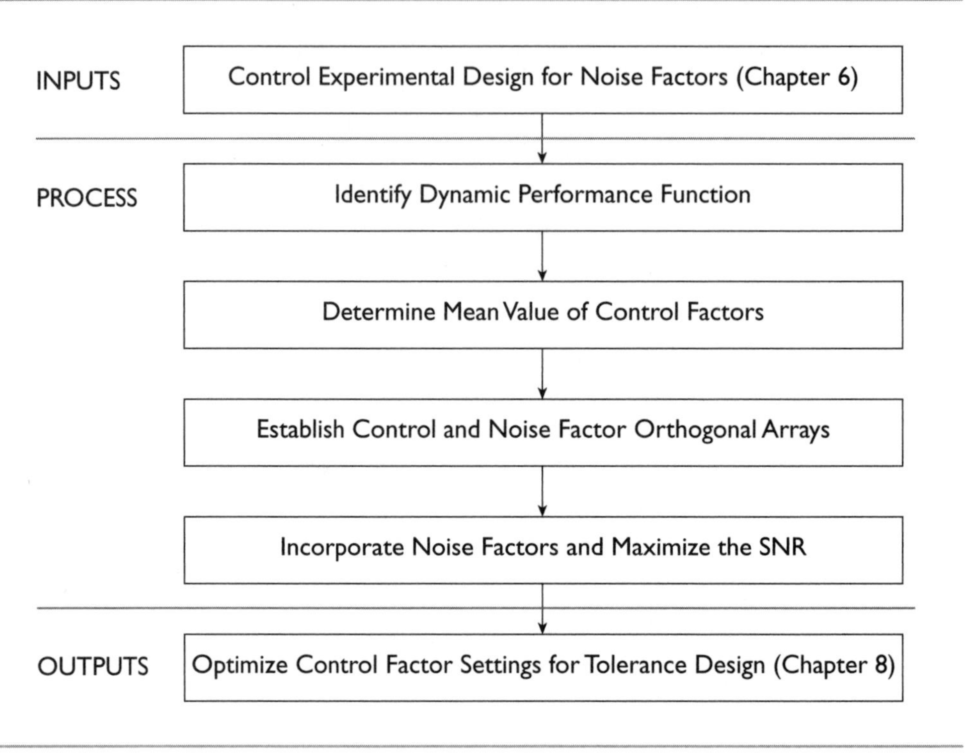

Figure 7-16 Process flow for a dynamic Parameter Design.

7.8 SUMMARY

Robust Parameter Design involves deciding the best values and/or levels for the control factors. The signal-to-noise ratio is a useful metric for that purpose. The robust design strategy entails the maximization of the SNR, which is the reduction of the noise with respect to the main factors' effects. For many engineering problems, the signal and response must follow a function called the ideal function. In the gyrocopter example in this chapter, the response (flight duration) and signal (wing length, and so on) follow a dynamic relationship. Such problems are called dynamic problems and the corresponding ratios are called dynamic SNRs, which are very useful for generating cost-effective engineering solutions to produce robust products and/or services.

BIBLIOGRAPHY

Box, G. E. P. "Signal-to-Noise Ratios, Performance Criteria, and Transformations." *Technometrics,* 30: 1–40, 1988.

Box, G. E. P., and S. Jones. "Split-plot Designs for Robust Product Experimentation." *Journal of Applied Statistics,* 19: 3–26, 1992.

Box, G. E. P., W. G. Hunter, and J. S. Hunter. *Statistics for Experimenters.* New York: John Wiley & Sons, 1978.

DeMates, J. "Dynamic Analysis of Injection Molding Using Taguchi Methods." In *Eighth Symposium on Taguchi Methods* (pp. 313–31). Dearborn, MI: American Supplier Institute, 1990.

Fowlkes, W., and C. M. Creveling. *Engineering Methods for Robust Product Design: Using Taguchi Methods in Technology and Product Development.* Englewood Cliffs, NJ: Prentice Hall PTR, 1995.

Gunter, B. "Discussion of Signal-to-Noise Ratios, Performance Criteria, and Transformations by G. E. P. Box." *Technometrics,* 30: 32–35, 1988.

Leon, R. V., A. C. Shoemaker, and R. N. Kacker. "Performance Measures Independent of Adjustment: An Explanation and Extension of Taguchi's Signal-to-Noise Ratio." *Technometrics,* 29: 253–85, 1987.

Mandel, J. *The Statistical Analysis of Experimental Data.* New York: Interscience, 1964.

Miller, A. E., and C. F. J. Wu. *Improving Calibration Systems Through Designed Experiments.* Research Report 9106, Institute for Improvement in Quality and Productivity, University of Waterloo, Waterloo, Ont., 1991.

Nair, V. N. (Ed.)."Taguchi's Parameter Design: A Panel Discussion." *Technometrics,* 34: 127–61, 1992.

Phadke, M. S. *Quality Engineering Using Robust Design.* Englewood Cliffs, NJ: Prentice Hall, 1989.

Pignatiello, J. J. Jr., and J. S. Ramberg. "Discussion of Off-line Quality Control, Parameter Design, and the Taguchi Method by R. N. Kacker." *Journal of Quality Technology,* 17(4): 198–206, 1985.

Shoemaker, A. C., K. L. Tsui, and C. F. J. Wu. "Economical Experimentation Methods for Robust Design." *Technometrics,* 33: 415–27, 1991.

Taguchi, G., S. Konishi, and Y. Wu (Eds.). *Taguchi Methods: Research and Development.* Quality Engineering Series, Vol. 1. Dearborn, MI: American Supplier Institute, 1992.

Taguchi, G., and Y. Yokoyama (Eds.). *Taguchi Methods: Design of Experiments.* Quality Engineering Series, Vol. 4. Dearborn, MI: American Supplier Institute, 1993.

Taguchi, S. "What Is Generic Function?" *Journal of Quality Engineering Forum,* 2(2): 40–42, 1994.

Welch, W. J., T. K. Yu, S. M. Kang, and J. Sacks. "Computer Experiments for Quality Control by Parameter Design." *Journal of Quality Technology,* 22: 15–22, 1990.

8 Tolerance Design

Minimizing Life-Cycle Cost

Designing for tolerance is the process of specifying allowed deviation from the nominal parameter settings that have been identified during the Parameter Design process. It involves balancing the added cost of tighter tolerances against the benefits to the customer because of reduced field failures. The Quadratic Loss Function—also known as the quality loss function (QLF)—is used to quantify the losses incurred by customers as a result of deviation from target performance.

As described in the previous chapter, the output of Parameter Design is the determination of nominal design values that maximize the robustness of a system under development. No assumptions are made concerning tolerances. During Tolerance Design, initial tolerance values are set based on the manufacturing capabilities and the cost targets for the design. Testing must then be conducted to determine whether the tolerances are appropriate; if they are, further detailed Tolerance Design is not needed.

Tolerance Design is third in the three-phase approach that Taguchi recommended for achieving robust designs. It needs to be done if Parameter

Design cannot produce the required performance without tightening process controls. The focus is on adjusting parameter tolerances at the points where variability could have a large negative affect on the final system. Similar decisions must be made with respect to the selection of different grades from available alternatives for subsystems and components. To be properly implemented, Tolerance Design involves a number of specific steps, including:

- Identifying the noises (tolerances) that will be investigated experimentally
- Setting the initial tolerance limits
- Formulating an experimental design for Tolerance Design experiments
- Analyzing methods appropriately to determine the individual contributions of each factor to be studied
- Determining the specific tightening or loosing of tolerance for each factor studied

As an engineering tool, Tolerance Design is a method for improving quality at an economical cost. It is used in conjunction with other tools and methods for the design of a product or process. This chapter starts with an introduction to some basic concepts related to Tolerance Design, including:

- The difference between Tolerance Design and tolerancing
- The relationship between component tolerances and system tolerances
- The cost and quality tradeoffs with which Tolerance Design deals
- The three phases of Dr. Taguchi's Robust Design Method

8.1 TOLERANCE DESIGN VERSUS TOLERANCING

Variation is a measure of the deviation in materials and dimensions from the nominal design goals. *Tolerances* are engineering specifications that allow for variation from the nominal, which is assumed to be the ideal setting. *Tolerancing* is a general term for the process of determining and communicating tolerances for systems, subsystems, and components. Tolerance Design is used for the following:

- Managing variability
- Achieving functional requirements
- Maintaining an economical cost structure

The specific Tolerance Design process determines component tolerances for a product, system, or service to deliver its intended function. Through designed experimentation, variability analysis, and loss function analysis, tolerances and the materials required to deliver functional requirements at the lowest overall cost are established.

EXAMPLE 8.1

The following electrostatic spray method is based on the principle that negatively charged objects are attracted to positively charged objects. Atomized paint droplets are charged at the tip of the spray gun by an electrode; the electrode runs 30kV to 140kV through the paint at 0 to 225 microamperes. Paint can be atomized using conventional air, airless, or rotary systems. The electrical force needed to guide paint particles to the work-piece is 8,000V to 10,000V per inch of air between the gun and a work-piece.

The part to be painted, which is attached to a grounded conveyor, is electrically neutral, and the charged paint droplets are attracted to that part. If the charge difference is strong enough, the paint particles normally fly past the part and reverse direction, coating the edges and back too. This effect is called *wraparound* and increases transfer efficiency. Electrostatic spray is used by most appliance manufacturers.

When painting a compact freezer, the thickness of the coating is a combination of electroplate (E-coat), primer, basecoat, and clear-coat. Each has different requirements, including film thickness (Figure 8-1). The paint shop process proceeds as follows:

- Pretreat by cleaning, coating with zinc phosphate, and neutralizing with chromic acid
- E-coat in dip tank and DI rinse
- Apply basecoat as an electrostatic spray (high cost)
- Apply clear-coat (low cost)

The capital investment for a new liquid electrostatic spray system consisting of a spray gun, two-gallon pressure pot, and hoses and fittings can range from

- ➢ Variability in manufacturing when painting
- ➢ Henry Ford: "You can have any color you want as long as it is black."
- ➢ Voice-of-Customer—affordable, reliable, available, individual mass transportation
- ➢ Black paint was the easiest color to apply and the quickest to dry in the early days of the twentieth century
- ➢ Stopping distance depends on mechanical, thermal, and control designs

Figure 8-1 Controlling paint thickness variation.

$4,900 to $7,500. The capital investment required for a new powder electrostatic coating spray system, including powder application equipment, powder booth, cleaning system, and bake oven, can range from $75,000 to $1,000,000.

The challenge of Tolerance Design for the painting process involves trying to control paint thickness throughout the process in order to deliver a finished job that is within tolerance. The tolerancing process involves reviewing, evaluating, and establishing cost-effective controls, as follows:

1. Examining the impact (on overall thickness) of variability in individual thickness
2. Determining the costs of controlling for thickness
3. Establishing individual tolerances to cost-effectively achieve an acceptable level of variability in the paint's overall thickness
4. Determining tolerances for individual thickness so that an acceptable level is possible
5. Communicating the tolerances
6. Implementing changes in the paint shop process to meet individual thickness tolerances

As variation in the quality characteristic increases, the number of dissatisfied customers also increases. To satisfy customers, the company must provide good-value, quality products. This requires engineers to establish nominal values and tolerances that balance manufacturing costs and customer expectations. As discussed in Chapter 6, a three-phase approach to robust design should be applied:

Phase 1: Concept Design determines the appropriate level of technology necessary to provide customers with product performance that meets their requirements. During this phase, it may be necessary to generate new concepts, ideas, methods, and so on. Here, engineers take new ideas and convert them into solutions.

Phase 2: Parameter Design takes the technology deemed appropriate during Concept Design and optimizes it for robustness, producing a system that functions consistently, as intended, throughout a range of customer usage conditions. It identifies optimal product or process parameter settings—the nominal values—that make performance less sensitive to noise. The design of experiment (DOE) process (using orthogonal arrays) is a major tool for achieving Parameter Design objectives.

Phase 3: Tolerance Design improves quality at an economical cost by selectively tightening or loosening tolerances, around the nominal values established during the Parameter Design phase, to reduce performance variation at minimal cost.

For maximum benefit at minimal cost, the three Robust Design phases should be applied in the order specified. While moving from Phase 1 to Phase 3, the cost of implementing modifications increases significantly. So, the DOE should maximize the benefit of each phase before moving on to the next one; in particular, Tolerance Design should be done after Parameter Design determines nominal values at which the system is most robust.

By using Parameter Design, system variability can be dramatically reduced using only low-cost alternatives to current parameter settings. If there are still too many dissatisfied customers when at the nominal values, then use Tolerance Design to further reduce response variation.

The Tolerance Design experimental design process is also used to determine the impact of component variability on system response variability.

Then the cost to reduce component variability is compared to the economic benefit of the reduction in system variability. Reducing variability or upgrading materials as indicated by Tolerance Design almost always increases product or process manufacturing costs.

8.2 Reduction in Variability

To meet component tolerances, the manufacturing variability of the product or the component must be maintained within certain levels. Thus, changes in tolerances should imply changes in manufacturing variation, which will cause variation in the system's Critical-to-Quality (CTQ) characteristics.

Figure 8-2 depicts a relationship between component (factor X_i) variation and system CTQ (measured by response Y) variation. For Example 8.1 in the previous section, the component factor X_i can be related to the thickness of the basecoat, while system CTQ variation can be related to total overall paint thickness. One aspect of Tolerance Design is determining the relationship between variation in the factor and variation in the system's function.

The curve on the Y-axis is the distribution of system CTQ (i.e., overall paint thickness) resulting from the variation in factor X_i (i.e., basecoat thickness). The system CTQ, Y, is a quality characteristic that measures system function.

In Tolerance Design, engineers seek factors for which a reduction in their variation results in a reduction in system CTQ variation (i.e., a reduction in system function variability). It involves deliberately varying the factors around their nominal values (in a designed experiment) to determine the percent of total system CTQ variation contributed by each factor's varia-

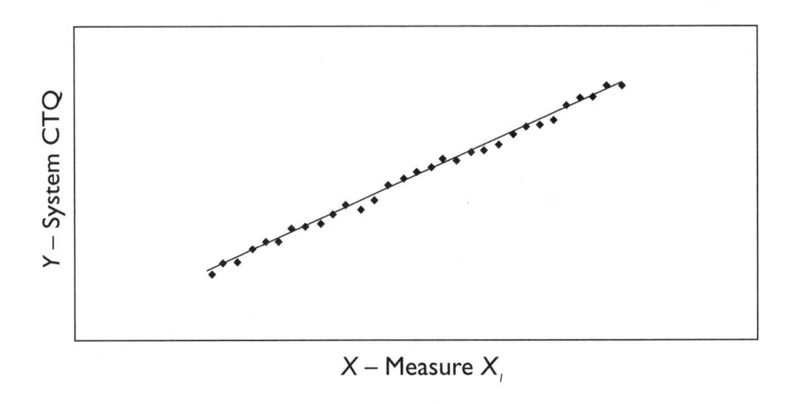

Figure 8-2 System CTQ response-versus-component variation for X_i.

tion. In some cases, reducing the variation of a factor (X_i) has little or no impact on system function (Y) variation.

In Example 8.1, it was necessary to determine the percentage of the paint's overall thickness that was contributed by the variation in the thickness of the basecoat. This was followed by a comparative analysis of the cost to reduce the variation (i.e., tighten paint thickness tolerance) and the cost of customer dissatisfaction at the present level of variation (i.e., dissatisfaction as a result of variation in total paint thickness).

To determine the effects of factor variability (around nominal values determined by Parameter Design) on system CTQ response variability, Tolerance Design uses an experimental design (see Chapter 7). Through designing for tolerance, factor tolerance adjustments can be prioritized based on each's impact and the cost to reduce such tolerances plus the cost implied by not reducing variability.

8.3 TIGHTENING TOLERANCE SELECTIVELY TO MAXIMIZE QUALITY-VERSUS-COST TRADEOFFS

A typical reaction to excess functional variation is to tighten the tolerances (reduce variability) of *all* of the components, which inevitably adds to manufacturing cost (see Figure 8-3). Since designs are likely to be more sensitive to some component variations than to others, a more sensible approach would be to tighten tolerances selectively.

Such selection depends not only on the relative impact of factor variability but also on the present level of economic loss associated with it plus the cost to reduce system variability. Doing so enables engineers to improve quality while limiting the impact on cost.

The relationship between quality and cost is used to rationally determine tolerances, resulting in improved quality, reduced costs, and increased communication between design and manufacturing disciplines. That is, the benefits of Tolerance Design are:

- Improvement in quality at the lowest cost by tightening tolerances or upgrading parts only whenever sensible

> ➢ Cost of labor to remove the failure mode
> ➢ Cost of extra materials
> ➢ Cost of extra utilities
> ➢ Cost of lost opportunities
> ➢ Loss of sales, revenue, and/or profit margin
> ➢ Loss of market share

Figure 8-3 Quality-versus-cost tradeoffs, including cost of poor quality.

- Reduction of costs by relaxing tolerances of some factors that have no effect on system variability
- Facilitation of design and manufacturing engineers' realization of common goals in order to develop a joint strategy

8.4 QUANTIFYING THE QUALITY LOSS FUNCTION

One of the key concepts in Tolerance Design is the quality loss function (QLF). This section covers the following:

- A description of the quality loss function
- The relationship between QLF and tightening tolerances
- The characteristics needed to determine the QLF for a specific system
- The average quality loss
- The economic benefits in terms of QLF

In Tolerance Design, the concept is used to weigh cost and quality tradeoffs. It provides an estimate of financial Loss to Society, $L(y)$, as a function of the deviation of the quality characteristic, y, from its target value, t.

The QLF philosophy is based on the fundamental assumption that any deviation in the quality characteristic from the target results in sub-optimal performance and that performance decreases as the deviation increases. Thus, if tightening component tolerances reduces variation in the quality characteristic, financial losses will be reduced. As shown in Figure 8-4, the quality loss function asserts that L (cost of customer dissatisfaction) increases in proportion to the square of the deviation of y (system CTQ) from t (target). In the figure, k is the loss coefficient.

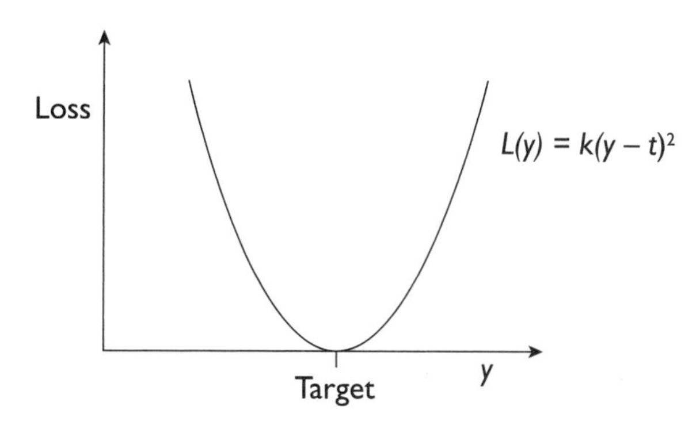

Figure 8-4 Quantifying a quality loss function.

This model is based on the assumption that there is no loss when *y* is on-target, and that the growth rate of *L* increases with the deviation of *y* from the target. This philosophy of loss represents a shift from the traditional (*goalpost*) perspective of quality; that is, anything within the specifications (represented by goalposts) is considered equally good, and loss is only incurred when the characteristic is out of spec.

Assuming that on-target performance, as shown in Figure 8-5, is represented by distributions with lower variation, such as distribution *a* versus distribution *b*, there is a higher percentage of systems close to the target. Therefore, a distribution with lower variation results in lower loss. Presumably, tightening component tolerances will require a reduction in the manufacturing variation of such components.

The challenge is to determine which tolerances to tighten to reap the most financial benefit. The area under the bell-shaped distributions represents the frequency of systems with the associated values for the quality characteristic—*y*.

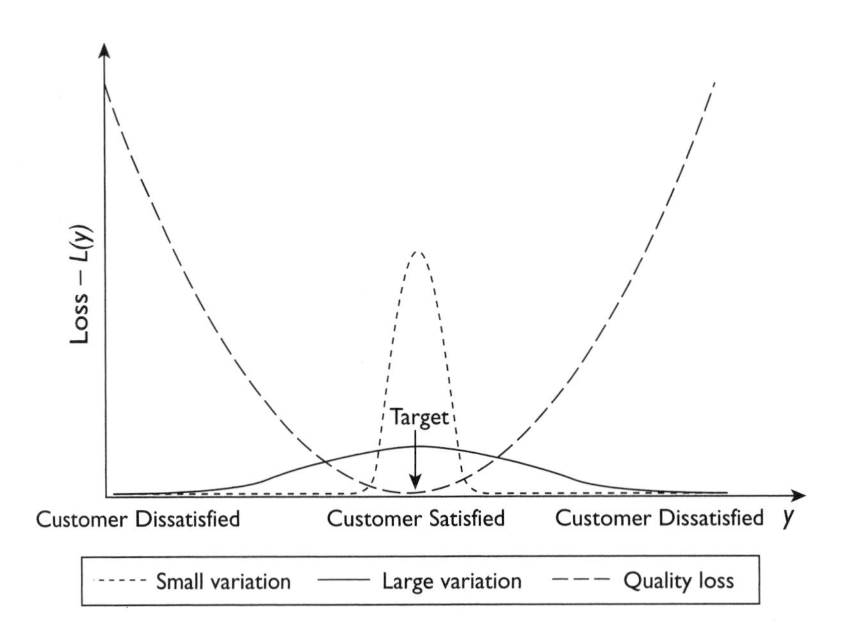

Figure 8-5 Satisfying customers by reducing variation.

The quality loss function shown in Figure 8-5 is for a nominal-the-best system CTQ; that is, a specific value is targeted for paint thickness and therefore total paint thickness is an example of this. The closer total paint thickness is to the design's nominal value, the higher customer satisfaction will be.

When engineering robust products with Six Sigma, engineers also deal with the smaller-the-better system CTQ (e.g., radio sound noise) or the larger-the-better system CTQ (e.g., measurement system accuracy). The QLF for a smaller-the-better system CTQ is:

$$L(y) = k_{y2};$$

and the QLF for a larger-the-better system CTQ is:

$$L(y) = k\frac{1}{y^2}$$

To estimate losses using the quadratic quality loss function, substitute the target value for t and estimate the k value. To estimate k, you must identify one point on the curve other than the vertex, which is assumed to occur at $(t, 0)$. It is usually easiest to estimate the point at which the *average* customer will return something for repair; this value of y (system CTQ) is denoted y_{50}, or LD_{50}.

As shown in Figure 8-6, y_{50} is the estimate of the point where the average customer returns for a repair. The loss at y_{50}, denoted here as $\$_{50}$, must

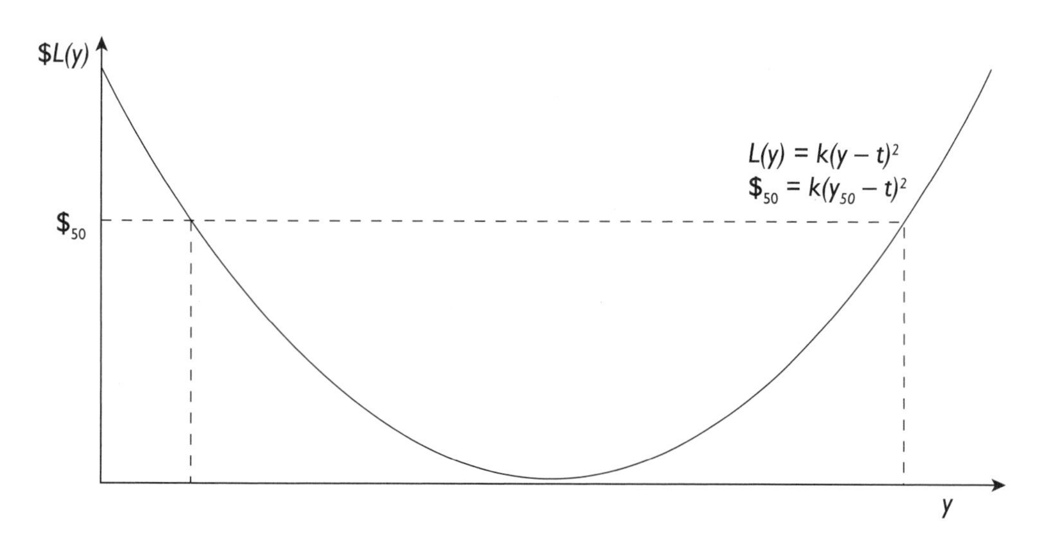

Figure 8-6 Determining a quality loss function.

also be estimated. The loss estimate must include all of the costs associated with the repair at y_{50}, plus the cost for parts and labor; any costs incurred by the consumer (e.g., missed work, alternative transportation, etc.); and the estimated cost of lost loyalty because of dissatisfaction. To find the loss function coefficient, *k*, substitute *t* and the estimates of y_{50} and $\$_{50}$ into the loss function and solve for *k*.

EXAMPLE 8.2

Suppose that for a particular system, $t = 10$mm, and estimates for y_{50} and $\$_{50}$ are 5mm and \$100, respectively; then the estimate of *k* for this system is:

$$k = \frac{100}{(5-10)^2} = \frac{100}{25} = 4$$

So, an estimate of the loss for this system at any value of *y* (system CTQ) can be found using its loss function, $L(y) = 4(y - 10)^2$.

As shown in Figure 8-7, related to the quality loss function is the average QLF, which estimates the average loss for samples with mean $\bar{y}$ and variance σ_y^2. This relationship is especially useful for cost–benefit analyses. The average quality loss is a function of the sample mean, $\bar{y}$, and sample variance, σ_y^2. Based on this formula, it is clear that loss can be lowered by either reducing σ_y^2, or by reducing the deviation of the mean from the target, $(\bar{y} - t)$. Frequently, engineers know how to shift the mean toward the target; so, to decrease loss, they must undertake the more difficult task of reducing σ_y^2. The average quality loss can be quantified by:

$$k\left[\sigma_y^2 + (\bar{y} - t)^2\right]$$

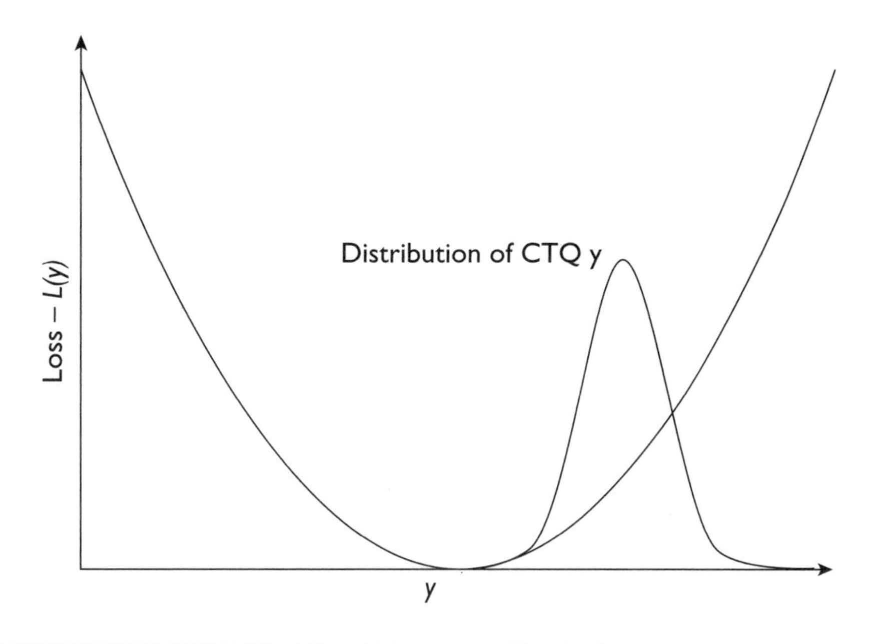

Figure 8-7 Average quality loss.

If system variation is reduced, quality loss will be reduced. The reduction in loss is an economic benefit. The average QLF can be used to estimate the economic benefit of reducing system variation. Provided that the mean is the same before and after reducing the variance, the benefit is directly related to the reduction in variance. Dr. Taguchi refers to the benefit as the *quality improvement.* The benefit of a new robust design can be quantified as:

$$
\begin{aligned}
Benefit &= \overline{L}_{current} - \overline{L}_{proposed} \\
&= k\left[\sigma^2_{current} + (\overline{y} - t)^2\right] - k\left[\sigma^2_{proposed} + (\overline{y} - t)^2\right] \\
&= k\left[\sigma^2_{current} - \sigma^2_{proposed}\right]
\end{aligned}
$$

Being able to reduce a product's sensitivity to changes in the signal is useful. For example, when designing a sports car, a desired outcome might be a car that allows the driver to change the feel of the road. Static Taguchi applications search for a product design, or a manufacturing process is used to average one fixed performance level. A static application for an injection-molding machine would be used to find the best operating conditions for a single-mold design. Dynamic applications use mold dimensions as the signal and search for operating conditions, which yield the same percentage of shrinkage for any dimension in any orientation.

The dynamic approach allows an organization to produce a design that satisfies today's requirements but can be easily changed to satisfy tomorrow's demands. You can consider this latter approach as contingency planning for some unknown future requirement. In dynamic applications, a signal factor moves the performance to some value and an adjustment factor modifies the design's sensitivity to this factor. If you plot a straight-line relationship, with the horizontal axis as the signal factor and the vertical axis as the response, the adjustment factor changes the slope of the line.

The signal factor would be a control knob setting. The analysis could determine that the suspension system is the adjustment factor, which adjusts the magnitude of change in road feel to a given change in knob setting. Several other design specifications would ensure a predictable relationship in the control knob setting and the feel. Changes in road conditions and weather would have minimal effect on the relationship between knob adjustment and road feel.

8.5 A Wheatstone Bridge System

A Wheatstone Bridge is a simple electronic circuit used to measure resistance, y (see Figure 8-8). The following example is used to illustrate Tolerance Design concepts. It shows how to select appropriately toleranced

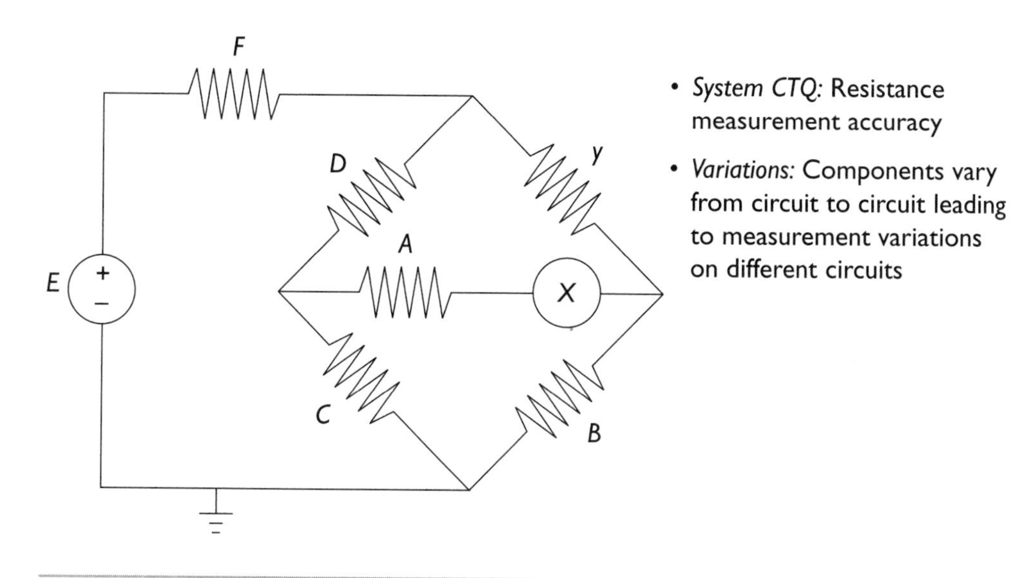
F
D
y
A
E
+
–
X
C
B
• *System CTQ:* Resistance measurement accuracy
• *Variations:* Components vary from circuit to circuit leading to measurement variations on different circuits

Figure 8-8 A Wheatstone Bridge is used to measure resistance, *y*.

components to minimize loss. Dr. Taguchi has used a similar example to describe how to apply his Tolerance Design methodology to electronic circuits' optimization (i.e., adjusting the system inputs to produce the best possible average response with minimum variability).

EXAMPLE 8.3

In this example, (1) the system response is the resistance measurement; (2) the system CTQ is the accuracy of *y*; and (3) the key design issue is how to discover nominal values of *A*, *C*, *D*, *E*, and *F* to minimize variability transmitted to *y*. Here, tolerances of the circuit components need to be tightened using Tolerance Design.

Based on the values—*A, B, C, D, E, F,* and *X*—of the circuit components, the measured resistance can be approximated using the formula:

$$y=\frac{B \bullet D}{C}-\frac{X}{C^2 \bullet E}\left[A(C+D)+D(B+C)\right]\left[B(C+D)+F(B+C)\right]$$

If the resistance values *A, B, C, D, E,* or *F* vary from nominal, the measurement of the unknown resistor will be affected. In addition, if there is current, *X*, flowing through the amp meter when it reads 0, then additional variation is introduced into the measurement.

Frequently, the reaction to quality issues of this sort is to tighten tolerances on all of the system's components. A more practical and recommended solution is to first use Parameter Design to determine component nominal values at which the system is less sensitive to variation. If measurement variation is not sufficiently reduced using Parameter Design, then employ Tolerance Design to determine the most sensible tolerances to tighten.

The next equation shows the calculation of *k* for the Wheatstone Bridge system; the repair and related costs are $\$_{50} = \$100$. An average customer returns a circuit when there is a 1 percent error in measurement; so, for a 2-ohm resistor:

$$k=\frac{\$_{50}}{(y_{50}-t)^2}=\frac{100}{(0.02)^2}=250{,}000$$

Using confirmation run data, y^i, for the optimal design from a Parameter Design experiment on the Wheatstone Bridge, the average quality loss (for this design) can be estimated. The value of $k = 250{,}000$ that is used here was calculated before. The average quality loss function can be calculated as follows:

$$\bar{L} = k\left[\sigma_y^2 + (\bar{y} - t)^2\right]$$
$$= 250{,}000\left[8.4 \times 10^{-5} + (2.0 - 2.0)^2\right]$$
$$= \$21$$

The average quality loss function includes the costs for replacement parts, labor, and lost customer loyalty. By comparing the average QLF for different tolerance specifications, Tolerance Design can help you choose tolerances on different parts of the design to minimize quality loss at minimal cost to the manufacturer.

In sum, quality loss function is a mathematical description of the relationship between economic losses to both the customer and the manufacturer and deviation of the system CTQ response from its target. y_{50}, or LD_{50}, is the response at which the average customer returns the system for repair. Economic benefit is the change in quality loss that results from a reduction in system variation.

8.6 Summary

When Parameter Design (see Chapter 7) is not sufficient for reducing the output variation, robust design's last phase is Tolerance Design. Narrower tolerance ranges must be specified for those design factors whose variation imparts a large negative influence on the output variation. To meet the tighter specifications, better and more expensive components and processes are usually needed. Because of this, Tolerance Design generally increases production and operating costs.

This chapter dealt with the problem of how, and when, to specify tightening tolerances for a product or a process so that quality and performance and/or productivity can be enhanced. Every product or process has a number—perhaps a large number—of components. How to identify the critical components to target when tolerances have to be tightened was explained here.

It is a natural impulse to believe that the quality and performance of any item can be improved easily by merely tightening up tolerance of some or all of its requirements. This means that if the old version of the item specified, say, machining to ±1 micron, engineers naturally believe that better performance can be obtained by specifying machining to ±½ micron.

Among the noise factors, if some have a much more significant affect on the system and result in a large variance in performance characteristics, you cannot achieve a combination of control factor levels that is insensitive to all of the noise factors. In this case, the noise factor that causes the most variation must be controlled by reducing it to obtain a lower variation—Tolerance Design, according to Taguchi's method. The first step is to determine the contribution of the noise factor to the variation—to identify which noise factors are causing a large variance. Next, consider ways of reducing the effect of the noise factor on the system. Through Tolerance Design, engineers can make appropriate economic tradeoffs between something's increased cost and its improved quality.

This can become expensive, however, and is often not a guarantee of much better performance. One merely has to witness the high initial and maintenance costs of such tight tolerance-level items (e.g., space vehicles, expensive automobiles, and so on) to realize that Tolerance Design (selection of critical tolerances and the respecification of them) is not a task to be undertaken without careful thought. In fact, it is recommended that only after extensive Parameter Design studies have been completed should Tolerance Design be performed as a last resort to improve quality and productivity.

It is difficult to make tradeoffs between product variation and components' cost, especially in practice, because most products have more than one quality characteristic. Thus, the objective of optimizing Taguchi's economic quality-selection model for Tolerance Design with multiple characteristics is to find optimal precision among components to minimize the total cost of variation and components.

As shown in Figure 8-9, Tolerance Design should follow system design and Parameter Design when products are unable to meet tolerance specifica-

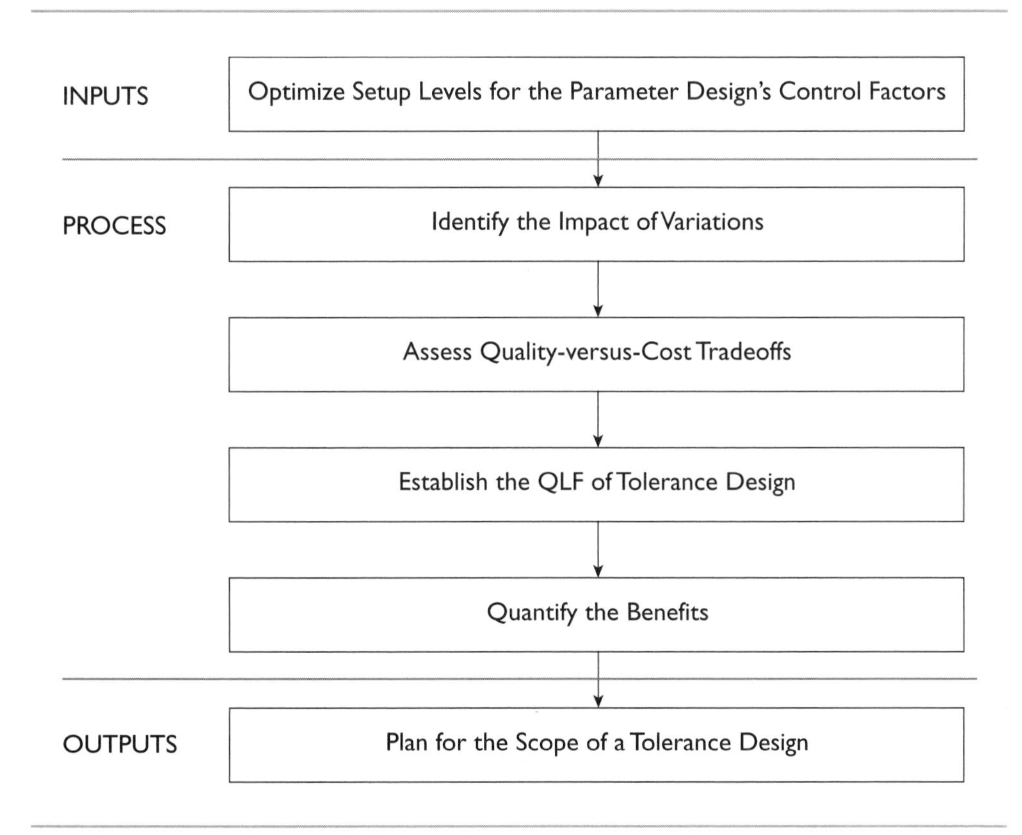

Figure 8-9 Tolerance Design planning.

tions or satisfy customers. Then, designers have to upgrade the precision of products and/or components to reduce losses caused by variations. This, in turn, increases production costs.

BIBLIOGRAPHY

American Supplier Institute. *Taguchi Methods: Implementation Manual.* Dearborn, MI: Author, 1989.

Bendell, A. "Introduction to Taguchi Methodology." *Taguchi Methods: Proceedings of the 1988 European Conference* (pp. 1–14). London: Elsevier Applied Science, 1998.

Bryne, D. M., and S. Taguchi. "The Taguchi Approach to Parameter Design." *ASQC Quality Congress Transactions* (p. 168), Anaheim, CA, 1986.

Creveling, C. M. *Tolerance Design: A Handbook for Developing Optimal Specifications.* Englewood Cliffs, NJ: Prentice Hall PTR, 1997.

Cullen J., and J. Hollingum. *Implementing Total Quality.* New York: Springer-Verlag, 1987.

Di Lorenzo, R. D. "The Monetary Loss Function—or Why We Need TQM." Paper presented at the Twelfth Annual Conference of the International Society of Parametric Analysts (pp. 16–24), San Diego, 1990.

Kacker, Raghun. "Off-Line Quality Control, Parameter Design, and the Taguchi Method." *Journal of Quality Technology,* 17(4): 176–88, 1985.

Logothetis N., and J. P. Salmon. "Tolerance Design and Analysis of Audio-Circuits." *Taguchi Methods: Proceedings of the 1988 European Conference* (pp. 161–75). London: Elsevier Applied Science, 1988.

Meisl, C. J. "Parametric Cost Analysis in the TQM Environment." Paper presented at the Twelfth Annual Conference of the International Society of Parametric Analysts (pp. 76–79), San Diego, 1990.

Phadke, M. S. *Quality Engineering Using Robust Design.* Englewood Cliffs, NJ: Prentice Hall, 1989.

Sullivan, L. P. "The Power of the Taguchi Methods." *Quality Progress,* June: 76–79, 1987.

Taguchi, G. *Introduction to Quality Engineering.* Asian Productivity Organization, 1986; distributed in the U.S. by American Supplier Institute, Dearborn, MI.

Taguchi, G., and S. Konishi. *Orthogonal Arrays and Linear Graphs.* Dearborn, MI: American Supplier Institute, 1987.

Teicholz, E., and J. N. Orr. *Computer-Integrated Manufacturing Handbook.* New York: McGraw-Hill, 1987.

Wille, R. "Landing Gear Weight Optimization Using Taguchi Analysis." Paper presented at the Forty-ninth Annual International Conference of the Society of Allied Weight Engineers. (pp. 161–75), Chandler, AZ, 1990.

9 Reliability Design

Giving Customers Long-Lasting Satisfaction

High reliability means a product has a long useful life and a high resale value, which gives customers long-lasting satisfaction. A product's useful life is driven by its strength against failure related to mechanisms of manufacturing and/or assembly, application environment, and wear-out. Reliability Design is a systematic approach to the design process that focuses on product reliability based on a thorough understanding of the *physics of failure*—an engineering term for understanding the root-cause failure mechanisms.

Reliability should be designed and built into products at the earliest possible stage of development. For a Six Sigma design, reliability can be ensured with five 9s: 99.99966 percent. As this chapter illustrates, reliability can be calculated based on the sigma levels determined by the interference between stress and strength. Here you will see how to establish a reliability plan,and how to predict reliability and sigma levels early in a product's life cycle.

9.1 RELIABILITY AND SIX SIGMA

Reliability is the probability that a product will perform its intended function for a specified time, under the various operating conditions encountered, in a manner that meets or exceeds customer expectations. It is an important aspect of engineering robust products with Six Sigma. Competition, warranty policies, ever-increasing customer expectations, and extended ownership increase the emphasis on reliability as a dimension of quality (see Figure 9-1).

Reliability focuses on the probability of maintaining an intended function over time. Once the customer has owned a vehicle for some time, the optional features, which may have been important to make the initial sale, are no longer as relevant or as important as other quality factors. Factors that enter into customers' perception of reliability and quality include useful life, signs of wear, and cost of maintenance. For example,

- ➢ Reliability is the probability that a product will perform its specified function adequately for a specified period of time under specified environmental conditions.
- ➢ Quality means:
  - Conforming to requirements and specifications
  - Fulfilling or exceeding the requirements or expectations of all internal and external customers
  - Meeting customers needs, as measured by the extent to which a product or service satisfies them

Figure 9-1 Reliability and customer-perceived quality.

the color of the interior becomes secondary to reliability over the life of a vehicle.

In sum, customers want products and/or services that meet or exceed their needs and expectations at a cost that represents value throughout something's service life. Reliability Design gives customers long-lasting satisfaction. Implied in the definition of reliability are three distinct viewpoints:

- Metric
- Customers' perception
- What engineers do

9.1.1 RELIABILITY AS A METRIC

The purpose of a metric is to quantify information. In the case of reliability, the metric is used to predict the probability of the success of a test. Of course, any test needs to be representative of customer usage. Usually engineers measure unreliability to determine reliability.

The following formulas demonstrate the relationship between reliability and unreliability. Sources of information for quantifying reliability are test data, fleet data, field data, and computer-aided engineering and/or hardware tests.

Relationship of reliability and unreliability:

$$\begin{aligned} R(t) &= \text{Probability of survival at time } t \\ &= 1 - [\text{Probability of failure at time } t] \\ &= 1 - \text{Unreliability} \end{aligned}$$

Reliability of a group of items tested:

$$\begin{aligned} R(t) &= \frac{\#\,Successes}{\#\,Tested} \\ &= 1 - \frac{\#\,Failures}{\#\,Tested} \\ &= 1 - Unreliability \end{aligned}$$

Notice the second formula; this value can be used as a relative measure for comparing designs. It is often referred as a *point estimate* of reliability. In the field, it may not be a good estimate of reliability because of the following:

- The unrepresentative nature of prototype parts tested
- Manufacturing variation of product parts
- Unaccounted for variation in customer usage, environment, and so on

For example, if information shows that after 10 years out of 100 alternators tested, 13 fail to function properly, the reliability of the group tested is 87 percent.

$$\begin{aligned} R(t) &= \frac{87}{100} \\ &= 1 - \frac{13}{100} \\ &= 87\% \end{aligned}$$

Here are some common unreliability measures:

- Percentage of failures in a total population—% failure
- Mean time between failure (MTBF)
- Mean time to failure (MTTF)
- Repairs per thousand (R/1,000)
- B_q life—the life at which the *q* percent of the population will fail

9.1.2 Customers' Perception of Reliability and Robust Design

Which words are used by customers to express reliability? The following would be among them:

- Well made
- Durable
- Dependable

In a technical sense, customers want a high level of function—robustness—over time, mileage, and usage cycles.

Robustness is the ability of a system to perform its intended function with low variability in the presence of noise, which causes variation in performance. Degradation of function is undesirable. A vehicle that becomes more difficult to start under certain conditions, say after 50,000 miles, is an example of degradation of function; the customer who defined reliability as "dependable" is no longer satisfied by the vehicle's reliability.

Customers are keeping their cars and other vehicles, and home electronics longer; thus, reliability needs to be designed in to meet reliability and quality expectations. Major vehicle and home electronics manufacturers, such as Toyota and Sony, have been able to differentiate themselves in the

- ➢ Minimize the impact of mechanical, thermal, and electrical stress and/or reduce sensitivity to them
- ➢ Design for the entire expected range of the operating environment
- ➢ Derate (limit the stress/load that can be applied to a component/system to levels below its upper limit) components and/or systems for added design margin
- ➢ Provide component and/or system redundancy
- ➢ Use parts and materials from suppliers known for offering quality components
- ➢ Reduce number of parts and interconnections needed
- ➢ Improve process capabilities for parts' manufacturing and assembly

Figure 9-2 Reliability Design—designed-in reliability and quality to meet customers' expectations.

marketplace because of their perceived reliability advantage (see Figure 9-2). This data says the following:

- Customer experience influences repurchase decisions of both first and second owners
- Reliability reputation affects resale value
- Improved reliability can win the loyalty of original and secondary owners

For example, a young customer may buy a seven- or eight-year-old vehicle. A bad experience with it could negatively influence the customer's decision to purchase a new vehicle of the same brand.

In addition, the perceived company/vehicle reliability reputation can affect the resale value/benefit, which is being advertised by competitors, to the original owner. The metric viewpoint of reliability is based on measurement. Reliability is related to the probability of failure in a test or in field usage. Feedback from customers says that product reliability is the most important to them.

9.1.3 What Is Done to Engineer Robust Products

Reliability can refer to the universe of things engineers do—that is, an *engineering process.* Reliability is part of their job. Everything that is done should support developing reliable products and processes; therefore, reliability is just good engineering.

The *reliability cycle* is a functional view of engineering activities. In reality, the timing of the activities actually overlaps. Based on the five Design for Six Sigma phases (see Figure 9-3), the reliability cycle is organized into sections, with each covering one of the following functional activities.

Define Reliability Requirements: The purpose of the first phase is to ensure that reliability requirements tie into customers' needs and/or wants and program objectives, with an emphasis on reusability.

Establish Reliability Metrics: The purpose of the second phase is to develop accurate and precise reliability measurement systems based on customer surveys. You cannot improve reliability if you cannot measure it.

Analyze and Diagnose Reliability: The purpose of the third phase is to ensure that there is a direct relationship between customers' performance requirements and the reliability requirements for new programs and to help maintain and improve the quality of present programs.

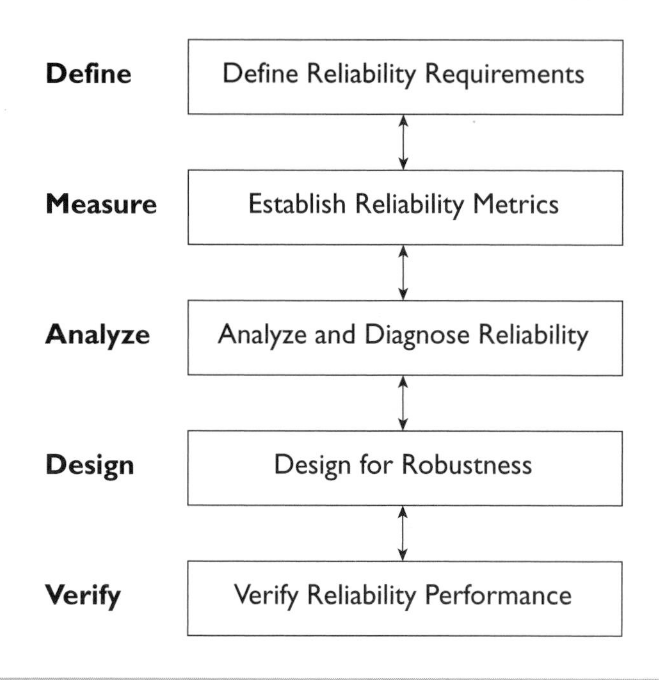

Figure 9-3 The five reliability phases.

Design for Robustness: The purpose of the fourth phase includes: (1) improving existing products' and processes' performance over time under customer-usage and manufacturing process conditions, and (2) inventing an inherently more robust new system or function.

Verify Reliability Performance: The purpose of the fifth phase is to demonstrate that functional targets are achieved. It also verifies the selected manufacturing process and its control plans by evaluating production runs, outlining mandatory production conditions, and identifying required outputs. Controls and relevant data are required to construct and determine control plan parameters.

Reliability Design emphasizes actions to prevent the occurrence of a potential nonconformance, and it stresses the importance of the control plan in the continuous improvement cycle. Engineers use two complementary strategies to achieve customer satisfaction: design-in function and design-out failure.

Design-In Function

This strategy reduces variability of the intended function in the presence of noise. Embodied in this approach is the Robust Method of Quality Engineering developed by Genichi Taguchi. Taguchi's method is to understand the ideal function of a system and to reduce response variation in the presence of customer-usage conditions (noise). The process to achieve this goal is to "engineer in" the intended (or ideal) function rather than "engineer out" problems. The robust method focuses on designed experiments using orthogonal arrays and consists of three phases:

1. Concept Design is the inventive phase.
2. Parameter Design is the phase during which the engineer determines the best design at the lowest cost based on a statistical explanation on the parameters. The experiments are designed to reduce variation by adjusting control factors' nominal levels.
3. Tolerance Design, the last phase, attempts to reduce variation by tightening tolerances. This is the most costly phase and is only used when the first two have not been able to achieve the robustness desired.

Design-Out Failure

This strategy's goal is to identify and minimize specific failures. Many traditional reliability engineering tools focus on this. Such tools have been well developed and applied in the automotive, aircraft, and defense industries. Although the traditional tools remain very useful in design, development, and manufacture of reliable products, an additional and complementary approach to reliability—reusability—is critical to engineers' continued success.

Reusability is a business strategy to deliver value in a competitive market. It is important to keep reusability in mind when designing and developing products, processes, and technologies.

- *Product*—use time-tested carryover systems, subsystems, and components
- *Process*—use established processes; make provisions in tooling and fixtures to accommodate future or derivative models without major tear-up
- *Technology*—use Parameter Design to develop knowledge of the robustness of commodities, subsystems, and so on that is applicable to a family of vehicles, not just a particular vehicle line; reuse knowledge about which control factors affect variability and which adjust the mean response

A r*oot cause* is an identified reason for the presence of a defect or problem; namely, the most basic reason, which if eliminated, would prevent recurrence. In engineering robust products with Six Sigma, *root-cause analysis* is the study of the original reason for nonconformance with a process. When the root cause is removed or corrected, nonconformance is eliminated.

EXAMPLE 9.1

Doris is a Design for Six Sigma Black Belt at a full-service supplier for laptop computers' OEMs (see Figure 9-4). A project team has been established to identify and to quantify the "field" environment in which a bracket is failing,

➢ What is the failure mode?
➢ What is the root cause for the failure?
➢ What is the current design control?
➢ Why did engineers not find the failure before?
➢ What corrective actions should be taken?
➢ How can the effectiveness of corrective actions be verified?
➢ What lessons can be learned from the failures?
➢ How can future failures be prevented?

Figure 9-4 Cracked or broken brackets on 4 out of 15 laptop computers.

and then to reproduce the failure in the original bracket design to confirm the failure mode. Based on the team's investigation, Doris receives the following information about the battery support brackets:

- Brackets develop cracks and eventually fail on some customers' laptop computers
- One commercial customer reported cracked or broken brackets on 4 of their 15 laptops
- Severe-duty customer usage for the battery support system is 80K cycles with a 0.17kg battery and 8g loading at 5Hz; support brackets had passed prior test requirements, which were less severe
- Stress pattern indicates a wide variation in stress level at different locations on the bracket

Concerning the different options for reliability improvement, the project team discussed the choice among the implementation approaches—add a spring, equalize or reduce stress levels through parameter redesign, or upgrade the material thickness. As a Black Belt, Doris needed to create an engineering plan to address the situation using the five Design for Six Sigma phases described in Section 9.1.3.

- Define reliability requirements—establish new performance requirements and develop a reliability engineering plan.
- Establish reliability metrics—determine reliability verification methods and which metrics to use and how the data is to be analyzed.
- Analyze and diagnose reliability—evaluate field performance. The way to implement a reliable engineering plan needs to include a thorough understanding of which strengths should be designed into the laptops' system to overcome stresses (loads) encountered during storage, shipping, and usage.
- Design for robustness—conduct analysis and experimentation and redesign the laptop brackets. Several design strategies can be used. For example, setting product parameters using the intersystem interactions increases the likelihood of the laptop exhibiting robustness (see Table 9.1).
- Verify reliability peformance—conduct verification testing of brackets under severe-duty usage; maintain quality in production using statistical process controls.

Table 9-1 Sample Applications: Design for Robustness

Design for Robust Ideas	Implementation Approach
Invent new design under the battery tray after a spring element or function	Add a spring element or use a new material (e.g., plastic)
Reduce variation of function	Equalize or reduce stress levels through Parameter Design
Minimize specific failures	Upgrade material thickness to eliminate cracking

9.2 Reliability and Stress–Strength Interference

To help understand the effects of variation on the product or process, engineers use the concept of stress–strength interference. This section looks at various interference scenarios and how the reliability of the product being studied is affected, including:

- Interpretation of stress–strength distributions
- Descriptions of five examples of stress–strength distributions

Both stress and strength are probabilistic rather than deterministic wherever stress and strength are distributions. Stress and strength in this context can be measured in any appropriate unit.

As shown in Figure 9-5, stress and strength both have normal distributions—the distribution shapes are equal in amplitude and spread. In real

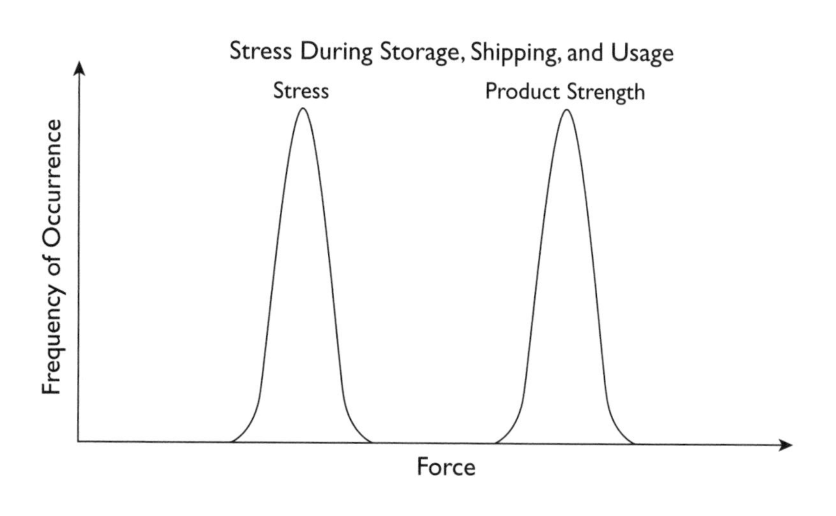

Figure 9-5 No or very little overlap between stress and strength.

- ➢ Static load
- ➢ Impact load
- ➢ Vibration
- ➢ Thermal conditions
- ➢ Electricity
- ➢ Electrostatic charging
- ➢ Sound waves
- ➢ Radioactivity

Figure 9-6 Various types of stress.

life, stress and strength are often very different. Notice in the figure that there is little or no overlap between the two, and it is rare for stress to exceed strength. When there is a lack of overlap between stress and strength, the component or system will be reliable.

Stress distribution represents the statistical distribution of overall stress severities among all of the customer-usage profiles, not the variation in stress amplitudes that may exist within one profile. Figure 9-6 lists some of the types of stress that could affect customer-use data.

Strength distribution can be influenced by various factors, such as lot-to-lot material variation and manufacturing variation from equipment settings, tooling, and so on. For example, the function of a radio speaker is to provide superior sound reproduction. The following are some stress and strength distributions for speakers:

- *Stress distribution*: There can be significant variation in stress, ranging from a customer who seldom listens to the radio or sets the sound volume low when in use (low stress) to a teenager who plays hard rock at a very high volume (high stress).

- *Strength distribution*: Variability in strength comes from differences in manufacturing materials, quality of connections, and so on.

9.2.1 Stress–Strength Distribution Overlap

Occasionally, the stress is greater than the strength, resulting in failure. The probability of failure for one application of stress can be calculated from the means and standard deviations of the distributions when they are normal. Engineered products, from concrete structures to electronic circuits, are designed to perform a function after selection of component part parameters that will permit successful operation in an expected use environment. Variation in part parameters, or in the operating environment, usually degrades the desired performance.

Figure 9-7 shows that variation in the strength of a part and variation in the stress it sees can result in an area of overlap in which the stress is

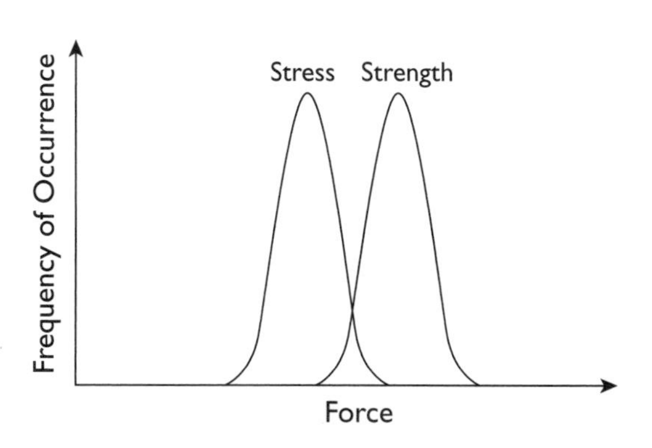

Figure 9-7 Stress–strength distributions overlap; failures occur when stress is greater than strength.

greater than the strength, resulting in a failure. For example, customer hand strength (stress) applied to a transmission shift lever may cause the plastic lever housing to break (strength).

EXAMPLE 9.2

The material used for interior trim in a minibus has been changed from a cool-polyester mix to nylon in order to improve wear characteristics. On vehicles built after the change, there have been complaints about shocks from static electricity, which is a performance problem. The horizontal axis for the load–strength curve is a combination of the amount of static electricity generated by the interior trim and its ability to carry away the electric charge from an occupant. Where there is an overlap, the occupant retains the charge when leaving the bus, then gets a shock on touching a ground. Variation between occupants cannot be reduced, and variation between seats is probably small. Improvement can be achieved by a design change that will remove the charge from the occupants. Any overlap between the probability distributions of stress and strength represents a probability of static (i.e., failure).

9.2.2 STRESS DISTRIBUTION WITH TAIL

If stresses beyond strength are applied, some units will eventually fail. The measured reliability would appear to have a constant failure rate if high stresses occur at random. The magnitude of the failure rate depends on the degree of overlap. In the next example, the design is not robust against occasional high levels of stress; things such as the following produce this distribution:

- A window-regulating electric motor system in a door; when the glass occasionally freezes in place, the motor is put under stress
- Supply voltage that occasionally produces surges

Example 9.3

Service centers report that they frequently need to tighten the bolts that hold the steering column in place. This problem occurs for cars of all ages. There is speculation among mechanics that some drivers cause the problem. This is a performance problem—the design is not adequate. The load distribution has a long tail that overlaps strength. High loads are probably caused by drivers who pull on the steering wheel to get in or out of the car, which results in forces that are much greater than what the designers expected. One solution is to increase the strength; another would be to provide a handle in a convenient location so that the steering wheel is not used for leverage.

9.2.3 Strength Distribution with Tail

All engineered products are subjected to common-cause variations, which are characterized by the following:

- Presence in every product or production process developed
- Produced by the process itself (the way business is done)
- Can be removed and/or lessened but requires a fundamental change in the process

Development of a product or a production process is stable, predictable, and in-control only when common-cause variation exists in the process.

Common-cause variation is fluctuation caused by unknown factors resulting in a steady but random distribution of output around the average of the data. It is a measure of the process potential, or how well the process can perform when special-cause variation is removed. In some cases, production units include a proportion of weak ones that will fail within a few applications of stress, but the majority continue to perform unaffected by stress. This strength distribution can result from a process that is in statistical control but not capable.

In other words, it is likely to result from common-cause variation, not special-cause variation. An example of this type of strength distribution is fasteners after they have been hardened and quenched; after that operation, the distribution will be long-tailed. Some weaker fasteners will fail under the load.

9.2.4 SUBPOPULATION WITH WEAK UNITS

In engineering product development, *population* is defined as the entire collection of items that is the focus of concern. Besides common-cause variations, products can be subjected to special-cause variations, which are characterized by the following:

- Unpredictability
- Typically large in comparison to common-cause variations
- Caused by unique disturbances or a series of them
- Can be removed or decreased by basic process control and monitoring

Development of a product or production process that exhibits special-cause variations is said to be out-of-control and unstable.

Occasionally, the tail of weak units forms into a second hump—a subpopulation. As illustrated in Figure 9-8, this distribution is bimodal and frequently occurs because of latent defects introduced during manufacture. This is most likely because of a special cause, which should be aggressively pursued and eliminated or contained. An example of a subpopulation of weak units is equipment malfunction that is a result of insufficient cure time for a molded seal, which can result in low-tear strength.

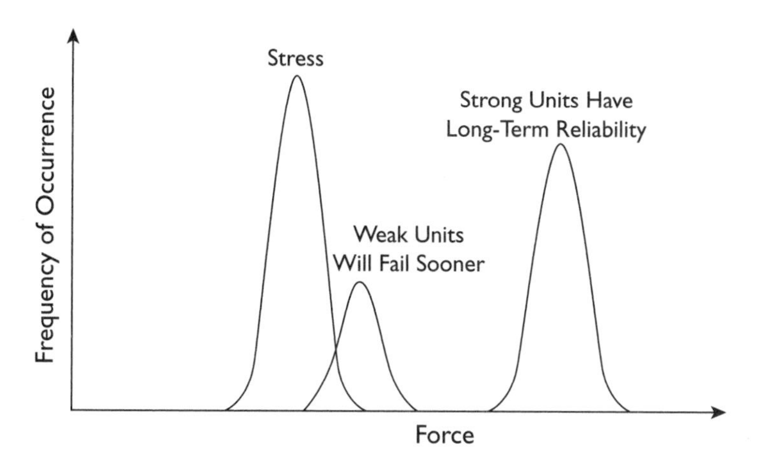

Figure 9-8 A subpopulation of weak units forms a second hump.

Example 9.4

For a first service call, about 1 in 20 owners report that carpets are lifting. This problem is rarely reported afterward. The cause is retaining clips that have worked loose. This is an "infant-mortality" problem—an example of common-cause variation. For the dealer, the remedial action consists of applying a sharp blow with a small hammer to the clips to ensure that they engage properly.

The strength distribution has a weak subpopulation (about 5 percent of the total) of retaining clips that have not been installed properly. Variation in assembly methods is the likely source of the problem. A design modification that makes the clips easier to engage and lock is the preferred solution. In the short term, though, a change to the process would be needed—push the clips harder, then tug on the carpet to check that the clips have been inserted properly.

9.2.5 STRENGTH DEGRADATION WITH AGING

The idea of engineering robust products with Six Sigma is to design a product that exhibits a state of robustness. Dr. Taguchi defines *robustness* as the state where technology, product, or process performance is minimally sensitive to factors causing variability (either in the production or user environment) and aging at the lowest manufacturing cost.

Most fairly accurate descriptions of equipment and/or process lifetimes assume that failure rates follow a three-period (i.e., I II III) "bathtub curve pattern"; that is, failures and/or errors:

I. Decrease during the debugging or improvement time period
II. Remain relatively constant and at their lowest levels during the normal equipment or process operating period
III. Increase during the wear-out time period

As shown in Figure 9-9, strength can deteriorate with time or use so that a previously immune population begins to fail. The mean can shift, and the variation in strength can widen because of wear or material degrading. Strength deterioration typically occurs in a limited-life component (e.g., a catalyst, deactivation; or tire, wear).

Note that stress can also increase over time. An example of this would be a lubricant whose protection of rotating parts degrades with age and allows increased stress on the components.

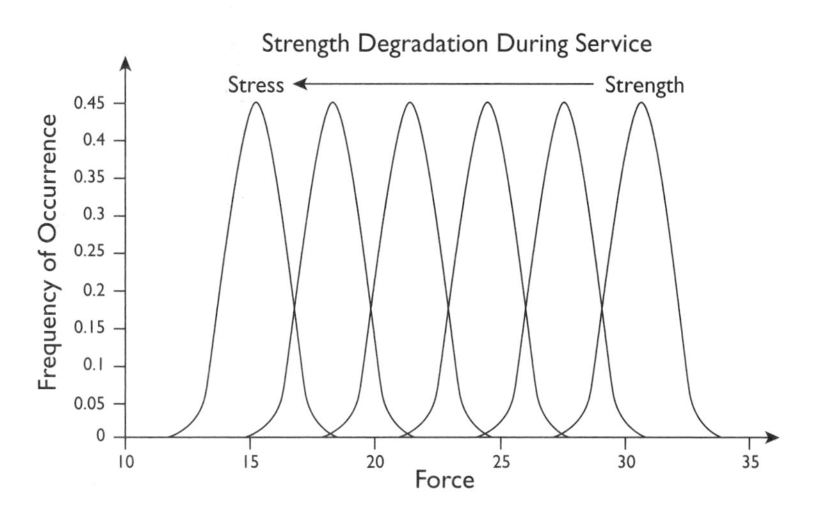

Figure 9-9 Strength degradation.

EXAMPLE 9.5

A rubber gaiter is fitted over the steering rack to prevent the ingress of dirt. It should last for the life of a vehicle, but some begin to show splitting between 45,000 and 55,000 miles. Frequently, gaiters are replaced at the 60,000-mile service. The strength distribution degrades with age; it begins to overlap with loads at 45,000 miles and is firmly into wear-out by 60,000 miles.

The horizontal axis for the load–strength plot is a measure of the functions of the gaiter. It has compound characteristics that can include flexing, resisting punctures, and preventing the ingress of fluids or solids at the seal joints. Part of the problem is in the cumulative damage experienced by the boots during use. If they could be made more robust, then the age at failure could be extended. Degradation of the material could also be slowed by making them more robust. The gaiters could also be made stronger at the start so that they can last for the required time at the current rate of wear. A more radical alternative is to change the design of the steering system so that there is no need to protect a rack.

The stress–strength interference method helps engineers understand the effects of variation on the product or process. Both the stress distribution and the strength distribution may be influenced by noise factors, which are the sources of variation that can affect these distributions.

As shown by Figure 9-10, the stress distribution is affected by outer noises, such as customer usage and duty cycle; the external environment; and the internal environment from neighboring subsystems. The deterio-

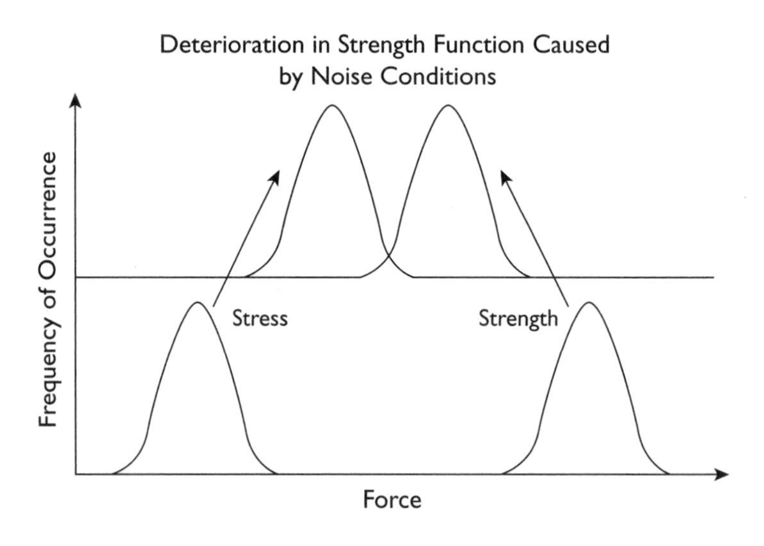

Figure 9-10 Stress (affected by outer noise) and strength (affected by inner noise) related to noise conditions.

ration in strength function is caused by noise conditions, such as piece-to-piece variation and aging, resulting in a change in dimension or strength over time and/or mileage.

9.3 Stress–Strength Interference Sigma Calculations

The failure of a single critical component in a modern complex system can have devastating consequences (e.g., failure of a disc in the turbine powering an aircraft). To ensure that such components are very unlikely to fail, a sigma calculation based on stress–strength interference is needed.

A simple approach uses a factor of safety, or design factor, that is in reality a factor of ignorance—to try and make an allowance based on experience with possible variations in materials, loading, and so on. This is not good enough for safety-critical items, for which some better quantification of reliability is needed.

A sigma calculation based on stress–strength interference uses probabilistic design, where the design variables that affect component strength are identified (e.g., variation in materials, geometry, surface condition, etc.) and their distributions, from which the distribution of the strength of the component is obtained. The distribution of factors affecting component stresses (e.g., load–stress concentration) are also determined. Once the two distributions have been obtained, component reliability can be calculated.

A product is broken (failed) when its stress exceeds its strength. If its strength keeps above its stress, the product continues to function. So, a stress–strength model, accordingly, deals with the probability that the product will keep functioning; that is, $R = P(X > Y)$—where X denotes the distribution of strength and Y denotes the distribution of stress.

Frequently, it is assumed that all of the distributions are normally distributed. The unreliability of the component is then obtained from the overlap of the distributions of strength and stress. One could determine the probability of failure based on the probability of stress exceeding strength. For example, use M to denote the separation of strength (H) and stress (S).

$$M = H - S$$

A product will not fail as long as M is positive (i.e., strength is greater than stress).

Given a distribution that describes the stress levels a unit will experience during its use and a distribution that describes the strength of the unit (e.g., material strength, material properties, design properties, etc.), the analyst can determine the probability of failure for this unit based on the probability of stress exceeding strength, or:

$$P[Stress \geq Strength] = \int_0^\infty f_{Strength}(x) \cdot R_{Stress}(x)dx$$

In this case, the data for the strength set would be actual data that is indicative of the strength of the material (i.e., maximum applied stress to cause failure), and the stress data would be actual data for the material under use conditions.

The relationship of stress, strength, and region of failure is shown graphically in Figure 9-11. This plot displays the probability density functions (pdfs) for the S and H distributions. A failure occurs when stress exceeds strength.

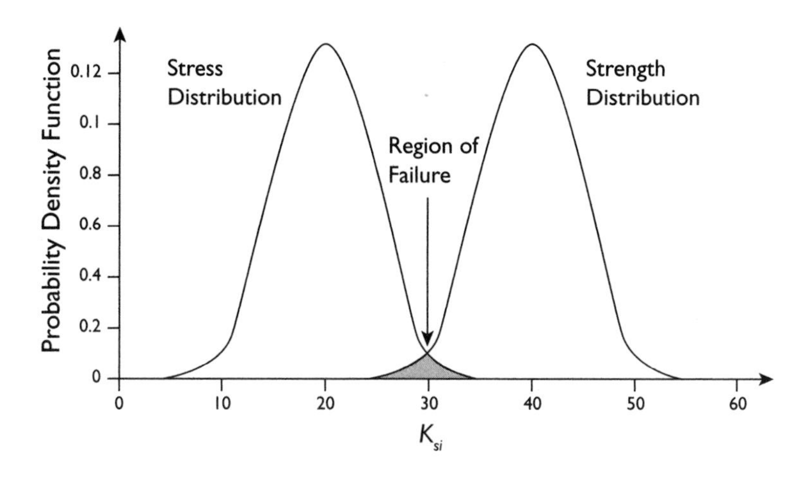

Figure 9-11 Stress distribution, strength distribution, and region of failure.

For certain distributions of strength and stress, simple expressions can be derived for reliability. This section considers the normal distribution, so strength and stress follow normal distributions:

$$N(\mu_H, \sigma_H)$$

and

$$N(\mu_S, \sigma_S),$$

where μ_H and σ_H denote the mean and standard deviation for strength, and μ_S and σ_S represent the mean and standard deviation for stress. Then, the reliability of the product can be expressed as:

$$R = P(H - S \geq 0) = P(X \geq 0),$$

where $X = H - S$ is the new random variable. Since X is a linear function of the normally distributed random variables *H* and *S*, it also follows that *X* is a normal distribution.

If strength and stress are assumed to be independent, the mean and standard deviation of *X* can be expressed as:

$$\mu_X = \mu_H - \mu_S$$

and

$$\sigma_X = \sqrt{\sigma_H^2 + \sigma_S^2},$$

where σ_x is the combined standard deviation of the strength and stress distributions. This measures the combined effect of variation in strength and variation in stress. Note that the form of the mean and standard deviation are useful in calculating the sigma value for reliability, as discussed next.

For many applications, there will be a lack of knowledge about loading and, consequently, the stressors. One of the reasons for improvements in the reliability of cars over the past thirty years is that automobile manufacturers have spent a lot of time measuring the effects on components of traveling over different surfaces.

This work has given vehicle designers a great deal of information about loads which, when combined with modern finite element analysis (FEA), enables them to predict mean and standard deviations of stress levels with a good degree of accuracy, which facilitates reliable design.

Design for Six Sigma is a methodology that provides engineers with the tools to improve the capability for engineering robust products. The objective is to reduce product output variation so that on a long-term basis, which is the customer's aggregate experience with processes over time, the result will be no more than 3.4 defect parts per million (PPM) opportunities (or 3.4 defects per million opportunities, DPMO).

For a process with only one specification limit (upper or lower), this results in six process standard deviations between the mean of the process and the customer's specification limit—hence, six sigma. Generally, engineers use Z_{lt}, called Z long term, to measure the sigma value.

For $X = H - S$, there is only a lower specification limit, which is 0. With its mean value, μ_x, and standard deviation, σ_x, the sigma value can be calculated as follows:

$$Z_{lt} = \frac{\mu_X}{\sigma_X} = \frac{\mu_H - \mu_S}{\sqrt{\sigma_H^2 + \sigma_S^2}}$$

Therefore, reliability can be determined by finding the value of the standard cumulative normal variable from the normal distribution tables, or an Excel function—NORMSDIST. The reliability can be calculated as follows:

$$R = P(X \geq 0) = \phi\left(\frac{\mu_H - \mu_S}{\sqrt{\sigma_H^2 + \sigma_S^2}}\right) = \phi(Z_{lt})$$

In Excel, assign the Z_{lt} to cell A1, then calculate reliability by:

= NORMSDIST(A1)

EXAMPLE 9.6

A component has a strength that is normally distributed, with a mean value of 4,200 N and a standard deviation of 368 N. The stress applied to it is also normally distributed, with a mean value of 2,780 N and a standard deviation of 412 N. The μ_X and σ_X can be calculated by:

$$\begin{aligned}\mu_X &= \mu_H - \mu_S \\ &= 4,200 - 2,780 \\ &= 1,420\end{aligned}$$

and

$$\begin{aligned}\sigma_X &= \sqrt{\sigma_H^2 + \sigma_S^2} \\ &= \sqrt{368^2 + 412^2} \\ &= 552.4\end{aligned}$$

Thus, the sigma value can be calculated as:

$$\begin{aligned}Z_{lt} &= \frac{\mu_X}{\sigma_X} \\ &= \frac{\mu_H - \mu_S}{\sqrt{\sigma_H^2 + \sigma_S^2}} \\ &= \frac{1,420}{552.4} \\ &= 2.57\end{aligned}$$

and the reliability is predicted as:

$$
\begin{aligned}
R &= \phi\left(\frac{\mu_H - \mu_S}{\sqrt{\sigma_H^2 + \sigma_S^2}}\right) \\
&= \phi(Z_{lt}) \\
&= \phi(2.57) \\
&= 99.49\%
\end{aligned}
$$

The Design for Six Sigma target for Z_{lt} is 4.5, which corresponds to 3.4 DPMO.

Using a normal distribution table, one can find that six sigma actually translates to about 2 DPBO. Normally, six sigma is defined as 3.4 DPMO, which really corresponds to a sigma value of 4.5. Where does this 1.5 sigma difference come from?

Motorola, which pioneered Design for Six Sigma, has determined through years of manufacturing–process data collection that manufacturing and assembly processes vary and drift over time; Motorola calls this the Long-Term Dynamic Mean Variation, which is typically around 1.5.

After a product has been improved using the Six Sigma methodology, the standard deviation and sigma value for stress and strength can be calculated. These are considered to be short-term values because the data contains only common-cause variations since product development projects and their associated collection of data occur over a period of 10 to 12 months rather than years. Long-term data, on the other hand, contains common-cause and special-cause (or assignable) variations. Because short-term data does not contain special-cause variations, it typically has a higher sigma level than long-term data. This difference is the 1.5 sigma shift. Given adequate product data, it is possible to determine the factor most appropriate for a specific process.

9.4 SUMMARY

Six Sigma Reliability Design is a mathematically based engineering design methodology for producing high-quality, mass-produced products. Six Sigma makes reliability calculations with the design parameters' probability distributions, instead of only the mean or nominal values. This allows the designer to design for a specific reliability or for a specific proportion of a product to be produced within specification—hence, guarantee safety, quality, and economy.

Incorporating Six Sigma reliability methods eventually leads to a better design approach in that the engineer develops a more comprehensive understanding of the problem that can encompass many disciplines. The reliability evaluation gives the designer an idea of the inherent risk, but, just as important, it provides a means of determining design parameter sensitivities. In general, Six Sigma Reliability Design methods require more detailed analysis, which ultimately leads to the design of improved, more efficient products, processes, and/or services.

BIBLIOGRAPHY

Aggarwal, K. *Reliability Engineering*. Boston: Kluwer/Academic Publishers, 1993.

Barlow, R. E., C. A. Clarotti, and F. Spizzichino (Eds.). *Reliability and Decision Making*. Norwell, MA: Chapman & Hall, 1993.

Brombacher, A. C. *Reliability by Design: CAE Techniques for Electronic Components and Systems.* New York: John Wiley & Sons, 1992.

Dovich, Robert A. *Reliability Statistics.* Milwaukee: ASQC Quality Press, 1990.

Ireson, W. Grant, and Clyde F. Coombs, Jr. *Handbook of Reliability Engineering and Management.* New York: McGraw-Hill, 1995.

Kapur, K. C., and L. R. Lamberson. *Reliability in Engineering Design.* New York: John Wiley & Sons, 1977.

Krishnamoorthi, K. S. *Reliability Methods for Engineers.* Milwaukee: ASQC Quality Press, 1992.

Misra, K. B. *Reliability Analysis and Prediction: A Methodology Oriented Treatment.* Oxford, UK: Elsevier Publishing Company, 1992.

Ramakumar, R. *Reliability Engineering: Fundamentals and Applications.* Englewood Cliffs, NJ: Prentice Hall, 1993.

Rao, Singiresu S. *Reliability-Based Design.* New York: McGraw-Hill, 1992.

Ushakov, Igor (Ed.). *Handbook of Reliability Engineering.* New York: John Wiley & Sons, 1994.

Wang, John X., and Marvin L. Roush. *What Every Engineer Should Know About Risk Engineering and Management.* New York: Marcel Dekker, 2000.

Wang, John X. *What Every Engineer Should Know About Decision Making Under Uncertainty.* New York: Marcel Dekker, 2002.

Zacks, Shelemyahu. *Introduction to Reliability Analysis: Probability Models and Statistics Methods.* New York: Springer-Verlag, 1992.

Appendix

The Process Map

Engineering Robust Products with Six Sigma

The process map for engineering robust products with Six Sigma is a hierarchical way of displaying processes that illustrates how a product is developed (Figure A-1). It is a visual representation of the workflow within the development process. Process mapping consists of a stream of activities that transforms a set of well-defined inputs into a predefined set of outputs.

Chapters 2, 3, and 4 discussed how to establish the Voice-of-Customer (VOC) models, and how to convert them into Critical-to-Quality characteristics (CTQs), design concepts, and design controls. The strategies to manage noises—referred to as white noise, random variation, common-cause variations, uncontrollable variables—were discussed in Chapters 5, 6, 7, 8, and 9. The following paragraphs briefly review these strategies.

Chapter 5: Engineering products present risks while providing benefits. Risk is about potential future failures when customers use existing products or engineers develop new products. As a predictive measure, the concept of risk is very important to minimize liability and maximize robustness.

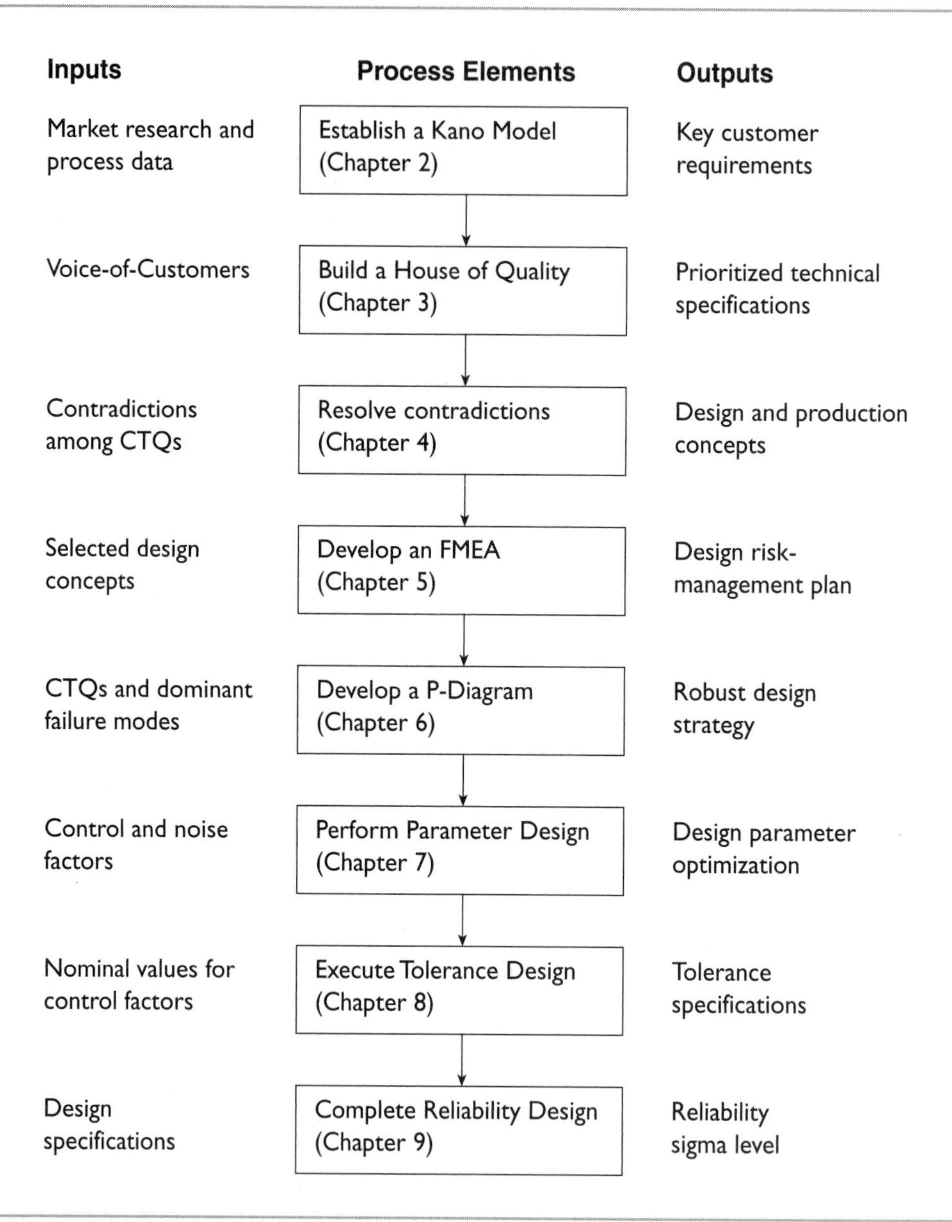
Inputs
Process Elements
Outputs
Market research and process data
Establish a Kano Model (Chapter 2)
Key customer requirements
Voice-of-Customers
Build a House of Quality (Chapter 3)
Prioritized technical specifications
Contradictions among CTQs
Resolve contradictions (Chapter 4)
Design and production concepts
Selected design concepts
Develop an FMEA (Chapter 5)
Design risk-management plan
CTQs and dominant failure modes
Develop a P-Diagram (Chapter 6)
Robust design strategy
Control and noise factors
Perform Parameter Design (Chapter 7)
Design parameter optimization
Nominal values for control factors
Execute Tolerance Design (Chapter 8)
Tolerance specifications
Design specifications
Complete Reliability Design (Chapter 9)
Reliability sigma level

Figure A-1 The process map for engineering robust products with Six Sigma.

Failure Mode and Effect Analysis (FMEA) helps to identify every possible failure mode of a product, process, or service, and to determine its affect on subitems and on something's required function. Through an FMEA, we can design controls to minimize the risk can be established. Design controls are the key elements for developing robust design strategies. This chapter illustrates how to develop an FMEA and how to use one to engineer for robust designs.

Chapter 6: To layout a robust design strategy, the product development team needs to identify the inputs and outputs associated with the design concept. Besides that, the team should understand what can be controlled cost-effectively and what is beyond its control. Parameters that can be specified by the designer are called control factors; those beyond the control of the designer are called noise factors. A large number of product failures and the resulting Cost of Poor Quality (COPQ) come from neglecting noise factors during the early design stages. Robust design involves deciding the best values or levels for control factors in the presence of noise factors. A P-Diagram, as described in this chapter, is an essential tool for every robust design project.

Chapter 7: Parameter Design is a principle that emphasizes the proper choice of levels for controllable factors in a process for the manufacturing of a product. When a design is said to be optimal, it implies that the design has achieved most of the target values set by quality measures before proceeding to a Tolerance Design. In an industrial setting, totally removing noise factors can be very expensive. Through Parameter Design, engineers try to reduce the variation around the target by adjusting the control factors' levels rather than by eliminating noise factors. By exploiting the nonlinearity of products and/or systems, Parameter Design achieves robustness, measured by a signal-to-noise ratio, at a minimum cost. Orthogonal arrays are used to collect dependable information about control factors with a small number of experiments.

Chapter 8: A Tolerance Design is the process of specifying allowed deviation of the parameters from the nominal settings identified during the Parameter Design process. It involves balancing the added cost of tighter tolerances against the benefits to the customer as a result of reduced field failures. The Quadratic Loss Function—also known as quality loss function (QLF)—is used to quantify the loss incurred by customers because of deviations from target performance.

Chapter 9: High reliability means a product has a long useful life and a high resale value, which gives customers' long-lasting satisfaction. A product's useful life is driven by its strength against failure mechanisms for manufacturing- and assembly-related failures, application environment-related failures, and wear-out failures. Reliability Design provides a systematic approach to the design process that is focused on product reliability based on a thorough understanding of the physics of failure. The physics of failure is an engineering term for understanding root-cause failure mechanisms. Reliability should be designed and built into products at the earliest possible stage of product development. For a Six Sigma design, reliability can be ensured with five 9s: 99.99966 percent reliable. As the chapter illustrates, reliability can be calculated based on sigma levels determined by the stress–strength interference.

Index

A

B

C

D

T

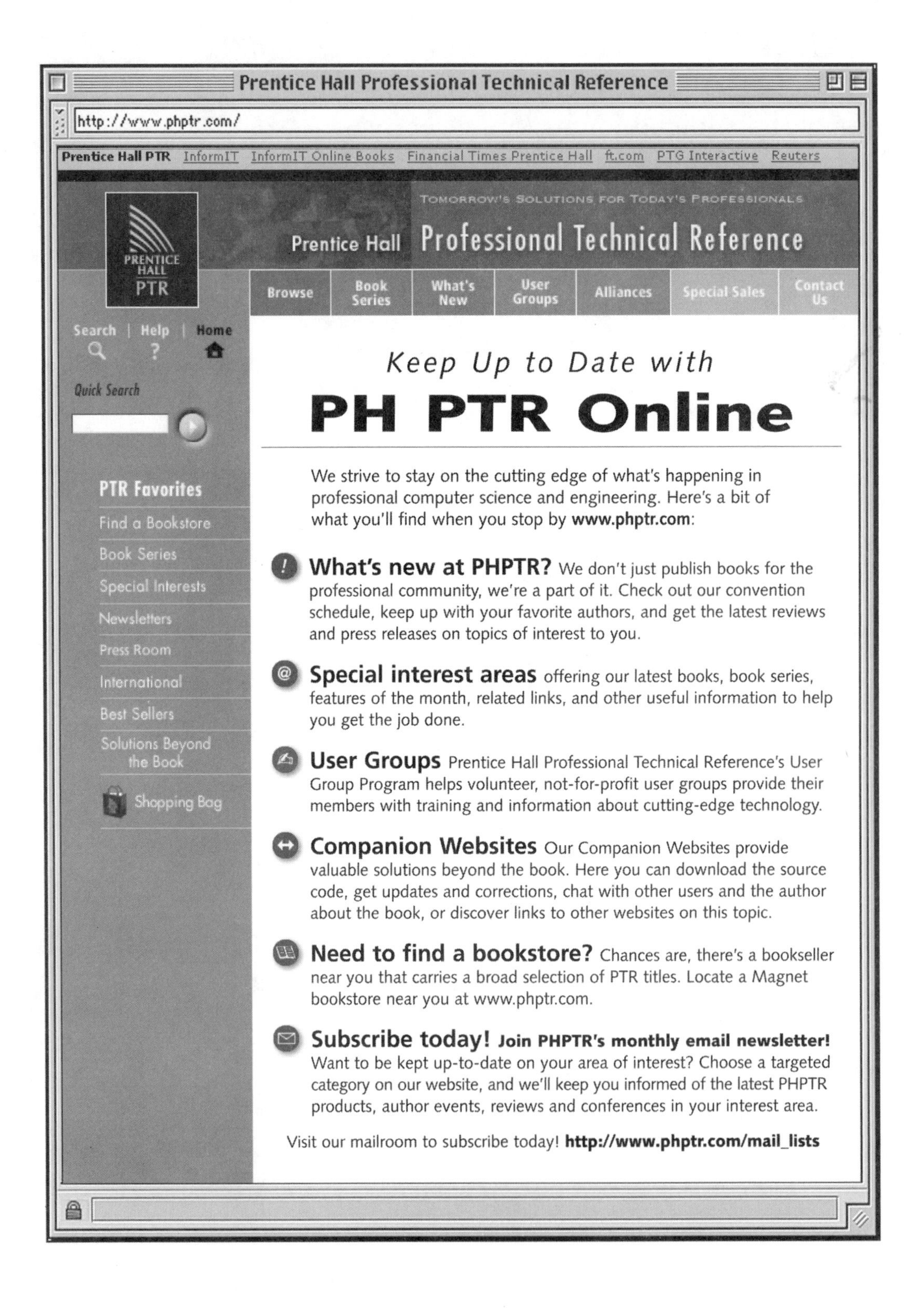

Prentice Hall Professional Technical Reference
http://www.phptr.com/
Prentice Hall PTR
InformIT
InformIT Online Books
Financial Times Prentice Hall
ft.com
PTG Interactive
Reuters
PRENTICE HALL PTR
TOMORROW'S SOLUTIONS FOR TODAY'S PROFESSIONALS
Prentice Hall Professional Technical Reference
Browse
Book Series
What's New
User Groups
Alliances
Special Sales
Contact Us
Search
Help
Home
Quick Search
PTR Favorites
Find a Bookstore
Book Series
Special Interests
Newsletters
Press Room
International
Best Sellers
Solutions Beyond the Book
Shopping Bag
Keep Up to Date with
PH PTR Online
We strive to stay on the cutting edge of what's happening in professional computer science and engineering. Here's a bit of what you'll find when you stop by www.phptr.com:
What's new at PHPTR? We don't just publish books for the professional community, we're a part of it. Check out our convention schedule, keep up with your favorite authors, and get the latest reviews and press releases on topics of interest to you.
Special interest areas offering our latest books, book series, features of the month, related links, and other useful information to help you get the job done.
User Groups Prentice Hall Professional Technical Reference's User Group Program helps volunteer, not-for-profit user groups provide their members with training and information about cutting-edge technology.
Companion Websites Our Companion Websites provide valuable solutions beyond the book. Here you can download the source code, get updates and corrections, chat with other users and the author about the book, or discover links to other websites on this topic.
Need to find a bookstore? Chances are, there's a bookseller near you that carries a broad selection of PTR titles. Locate a Magnet bookstore near you at www.phptr.com.
Subscribe today! Join PHPTR's monthly email newsletter!
Want to be kept up-to-date on your area of interest? Choose a targeted category on our website, and we'll keep you informed of the latest PHPTR products, author events, reviews and conferences in your interest area.
Visit our mailroom to subscribe today! http://www.phptr.com/mail_lists